Mentoring and Sponsoring

Maria Angela Capello · Eve Sprunt

Mentoring and Sponsoring

Keys to Success

Maria Angela Capello
Houston, TX, USA

Eve Sprunt
Dublin, CA, USA

ISBN 978-3-030-59432-9 ISBN 978-3-030-59433-6 (eBook)
https://doi.org/10.1007/978-3-030-59433-6

This Springer imprint is published by the registered company Springer Nature Switzerland AG
The registered company address is: Gewerbestrasse 11, 6330 Cham, Switzerland

For Alessandra and Claudia,
Maria Angela Capello

For Emmett,
Eve Sprunt

Foreword

This book on mentoring and sponsoring is a wonderful compilation of interviews from leaders in positions of influence. Their professions vary from engineers, geoscientists to a highly regarded tennis coach to sociologists to orchestra conductor and academics. Every one in this seemingly eclectic group had a genuine enthusiasm for helping young professionals. Regardless of the interviewee's profession and definition of mentoring and sponsoring, I found commonality in every interview. The overarching theme is that every person was genuinely passionate about caring for people, with respect and trust being key ingredients to a successful relationship. Whether these interactions were formal or informal, I found that each had humility and humor when speaking about their mentor and sponsor roles. It is interesting that many did not realize until they were advanced in their careers or maybe even during the interview that they had mentors and sponsors or that they were mentors and sponsors.

We on the more senior side of our careers will gain insight into the young professionals we mentor or sponsor in the last sections of the book—The other side of the coin: the mentees. Ask yourself the question: Is our mentoring and sponsoring style relevant to more junior professionals? For those interviewed in this book, I say a resounding YES!

The format makes reading this book delightful! You could spend hours reading the views of these interesting people or take 15 min to focus on one interview at the beginning or end of your day for inspiration.

I offer my personal thanks to Eve and Maria Angela for finding these exceptional professionals to interview and their thoughtful questions and analysis of these interviews. I consider both of these talented women (one a peer, the other younger) friends, mentors, and sponsors. I find joy being with them, I try and emulate their passion for life, and they have both given me or pushed me toward opportunities I would not have sought without their encouragement. I am inspired by you, I was inspired by the people in this book, and I learned something from each of the chapters.

I hope as you read these marvelous interviews, you also find something to help guide you in your personal and professional life in being a mentor or a mentee and being a sponsor or the one who is sponsored!

Dr. Ramona M. Graves
Professor Emeritus
Petroleum Engineering, Colorado School of Mines
Colorado, USA

Preface

We met in Houston in 2005, when we both were attending an annual meeting of the Society of Petroleum Engineers. We quickly established a collaboration that resulted in many pioneering initiatives for professional societies, especially the Society of Exploration Geophysicists (SEG) and the Society of Petroleum Engineers (SPE). We realized that besides excelling in our respective disciplines, successful career benefits from mentors and sponsors.

We wanted to share our insights on these two important career advancement enhancers, and a book seemed the right medium to do it. We expanded our learnings by including the visions of leaders and young professionals who generously shared their own perspectives and anecdotes on mentoring and sponsoring experiences that shaped their careers. We are overwhelmed with gratitude to every one of them, as now we can offer our readers a rich compilation of wonderful advice and practical tips that craft a learning platform of immense value.

The first interview was conducted in Rome, Italy, May 2019. After several months, there was a moment when we thought we would not finish our project. Then, March 2020 brought the global COVID19 Pandemic, and we used the stay-at-home period as an unexpected opportunity to finalize the interview process and focus on writing until all chapters were completed. The last interview was conducted in Kuwait in April 2020. We can proudly say we transformed a crisis into an opportunity that we now offer to you in the form of a book.

Houston, TX, USA Maria Angela Capello

Dublin, CA, USA Eve Sprunt
May 2020

Acknowledgements

We learned a great deal from many people, and we are grateful to those who offered us the right key at the right moment, as mentors or sponsors.

This book is not only the result of our efforts as authors but also the gifts of time and insights by the leaders and young professionals we interviewed. We want to start by thanking them for their generous assistance.

Carol Susan Howes skillfully edited all of the chapters, provided recommendations for improvement and was a most patient and frequent sounding board. We owe her our sincerest gratitude.

Dr. Herminio Passalacqua provided photos of the keys we used in this book. It is said that a picture is worth a thousand words. These pictures of keys show how the ability to open the lock in a door depends on subtle differences.

Dr. Alexis Vizcaino believed in our project from the beginning and pushed us to expand our scope beyond the energy sector, greatly enriching our compilation.

We reached out to leaders and young professionals through mutual connections. We especially want to recognize and thank Charlotte Griffiths, Alberto Bitossi, Olga Bravo, Yasmine Samarani, Paola Capello, Patrizia Capello, and Dr. Nansen Saleri. They have our utmost gratitude.

Contents

About the Authors

Maria Angela Capello is a renowned leader and author in the energy sector, and an expert consultant and advisor in reservoir management, sustainability, talent development, and diversity and inclusion. She was the first woman to supervise seismic crews in the jungles of Venezuela and has had a career that spans three continents.

She was an executive consultant at Kuwait Oil Company, has published 87 technical articles, and is the lead author of "Learned in the Trenches—Insights on Leadership and Resilience" (Springer). She is Knight of the Order of the Star of Italy (Cavaliere dell'Ordine della Stella d'Italia) and co-chairs the UN Women in Resources Management Committee. She received the highest individual recognition by the SPE, the Honorary Membership Award. She also received the SEG Special Commendation and Lifetime Membership Awards.

She is a Distinguished Lecturer of the SPE, chairs the SPE Business, Management and Leadership Committee, and serves as SEG Director at Large. Her passion is the empowerment of women and young in the sustainability of energy. She has an M.Sc. from the Colorado School of Mines.

Eve Sprunt was the 2006 SPE President. In 2010, she received SPE's highest recognition, Honorary Membership. She has 35 years of experience working for major oil companies, 21 years with Mobil, and 14 years with Chevron. In 2013, she received the Society of Women Engineers' Achievement Award. She was the 2018 President of the American Geosciences Institute. She was the founder of the Society of Core Analysts in 1985.

Her S.B. and S.M. degrees are from MIT and her Ph.D. (Geophysics) from Stanford. She was the first woman to receive a Ph.D. in Geophysics from Stanford.

She has authored more than 120 editorial columns on industry trends, technology, and workforce issues, 23 patents, and 28 technical publications. She is the author of A Guide for Dual-Career Couples, which was published by Praeger in May 2016. Her second book, Dearest Audrey: An Unlikely Love Story, about her aunt's adventures in Pakistan in the 1950s was published in 2019. She speaks and consults on both energy issues and women's issues.

Introduction

Our Goal

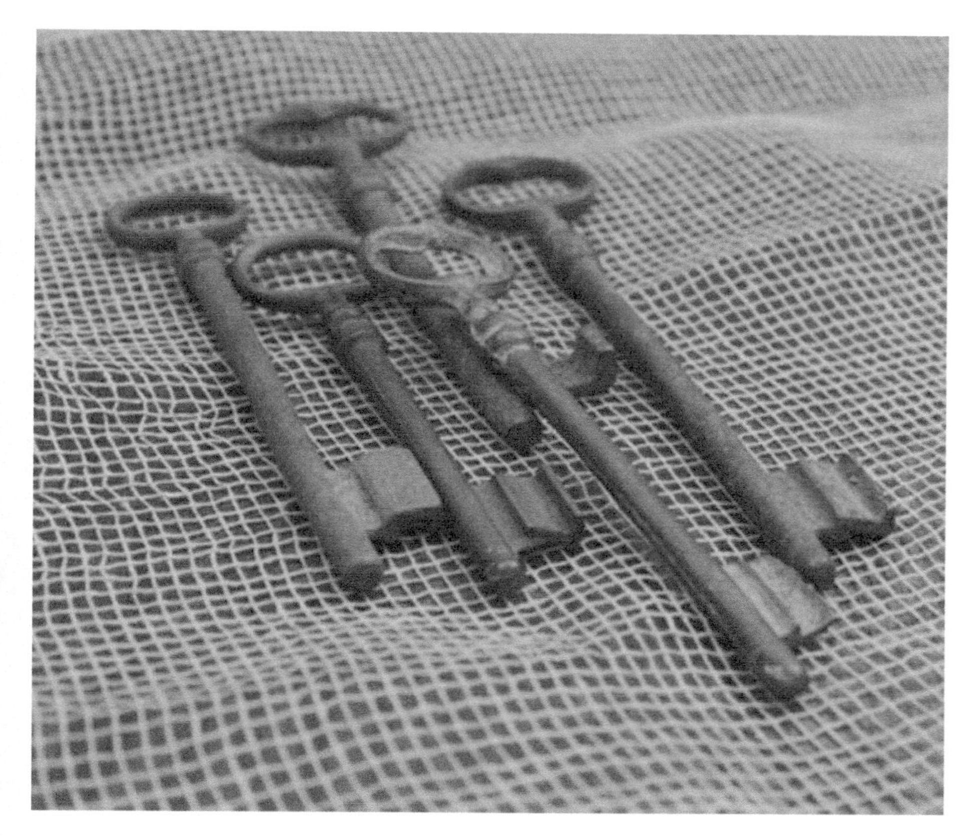

M. A. Capello and E. Sprunt, *Mentoring and Sponsoring*,
https://doi.org/10.1007/978-3-030-59433-6_1

Our goal is to provide valuable insights and lessons learned about mentoring and sponsorship to ambitious early and mid-career people. People who are mentored and have the right sponsor are more likely to fulfill their dreams and achieve success in their chosen careers. Mentoring can help us identify important skills to add to our repertoire and network more effectively. Sponsors can improve our chances of receiving advantageous job opportunities. Don't let time slip away from you. Find out what you need to know to succeed.

Technical preparedness and networking get us started, but mentoring and sponsoring help us stand out in comparison with our peers and are now essential for ambitious professionals. Even the most highly skilled technical professionals will find it difficult to rapidly advance and reach the highest ranks in their organizations without wise mentors and strong sponsors.

We have supplemented our own many years of experience by interviewing senior leaders from around the world. In the following chapters we share the wisdom these leaders have gained from their own careers and experiences as mentors and mentees. We also interviewed rising young stars to get their perspective on what they have found to be most beneficial.

Many of us tend to be somewhat shy about asking someone we don't know well for guidance. We may also be reluctant to share highly ambitious career aspirations and/or may not be sure of what we want. Asking knowledgeable more experienced people about their careers and our hopes and goals is a good way to build a plan. If you don't know where you want to go, your odds of getting there are much lower.

Explore the pages ahead either in consecutive or random order. Benefit from the experiences of the people we interviewed. These leaders and rising stars have opened their hearts and shared their memories with us with the goal of helping you gain perspective from their experiences! We invite you to engage in the same journey we took in learning about their rewarding and diverse experiences with mentoring and sponsoring.

We want our book to be a journey of discovery. We hope to empower you with the combined wisdom of this diverse group of leaders and a select group of promising young professionals.

The Challenges

Why a book on Mentoring and Sponsoring?

Our academic training does not fully prepare us to succeed at work. Yes, there is competition amongst students, but the focus is on technical knowledge rather than emotional intelligence, inter-personal skills and organizational politics. The top students in science, engineering, and any other discipline, do not necessarily reach the highest ranks of corporations or the highest technical leadership roles. They do not necessarily succeed as self-made entrepreneurs at the heart of new companies. In other words, 4.0 GPA cum-laude multi-degree graduates are not necessarily the leading candidates in a corporate succession-planning list and would not necessarily be the first choice of their own peers to lead their team. In life, soft skills, sponsoring and mentorship make a world of difference in career success.

The transition from being a student to an employee is a big shift. As students we focused on learning what was needed to ace tests in our selected disciplines to achieve our goal of earning a Bachelor of Science or Arts, and then a Masters, or a doctoral degree in our chosen field of knowledge. And we do this in systems which are designed to minimize subjective influences and preferences that would be considered unfair in an unbiased world. But those elements not only exist but play large roles in the real world. When we enter the work environment, we may lack necessary skills and/or awareness of which factors may impact our chances of technical and managerial success. We need to acquire business savvy and develop new strategies.

During our school years, we are judged based on standardized tests and evaluation systems that are supposed to be fair and equally applied to all. Learning that this is not how real-life works may take years. By the time some people recognize that other elements are equally or perhaps even more important to their career success than technical competency, it may be too late. They may already have been eliminated from succession plans for leadership roles.

Promotion processes in organizations are often dependent on factors such as:

- Key Performance Indicators (KPI)
- Ability to liaise with peers and supervise personnel
- Ability to liaise with their own supervisors or leaders
- The "perception" of the potential of the individual by the leaders/evaluators.

Many of these factors could be standardized and related to specific results and measurable factors. But, many forms of bias may influence the outcome. Some individuals have a higher profile with the decision makers and advance or secure a better lateral position.

Success depends on more than technical performance in private, governmental, and not-for-profit organizations and in academia. Soft skills, mentoring and sponsoring all have a significant impact. In this book we focus on the importance of mentoring and sponsoring, because they shape and propel careers.

Who was the First Mentor?

The words "mentor" and "mentoring" can be traced back to "The Odyssey," written by Homer. When Odysseus departs for Troy, he leaves his son, Telemachus in the care of his mother, Penelope, and two trusted friends and old advisors, his foster-brother, Eumaeus, and Mentor. Telemachus was a young child when his father sailed away to Troy. Odysseus's absence lasted almost twenty years. Telemachus grew up and had to manage many issues during his father's absence. Disguised as Mentor, the goddess Pallas-Athena advised Telemachus on several occasions including suggesting that Telemachus should expel his mother's suitors from the house.

In 1699, the French author, Francois de Salignac de la Mothe-Fenelon wrote the didactic French novel *Les adventures de Télémaque (The adventures of Telemachus),* in which he explains the importance of Mentor in the upbringing of Telemachus and initiated the modern usage of "mentor" and "mentoring."

Since Mentor was the female goddess Pallas-Athena, the advisor of Telemachus was a woman. However, the words "mentor" and the act of "mentoring" are now well understood independent of the specifics of the story as an advisor and wise coach for career development.

The Oxford English Dictionary defines "mentor" as "one who fulfils the office which the supposed Mentor fulfilled towards Telemachus. Hence, as common noun: An experienced and trusted adviser."

Merriam-Webster's Dictionary provides two definitions, "a friend of Odysseus entrusted with the education of Odysseus' son Telemachus" and "a trusted counselor or guide, tutor, coach."

Mentoring and Sponsoring

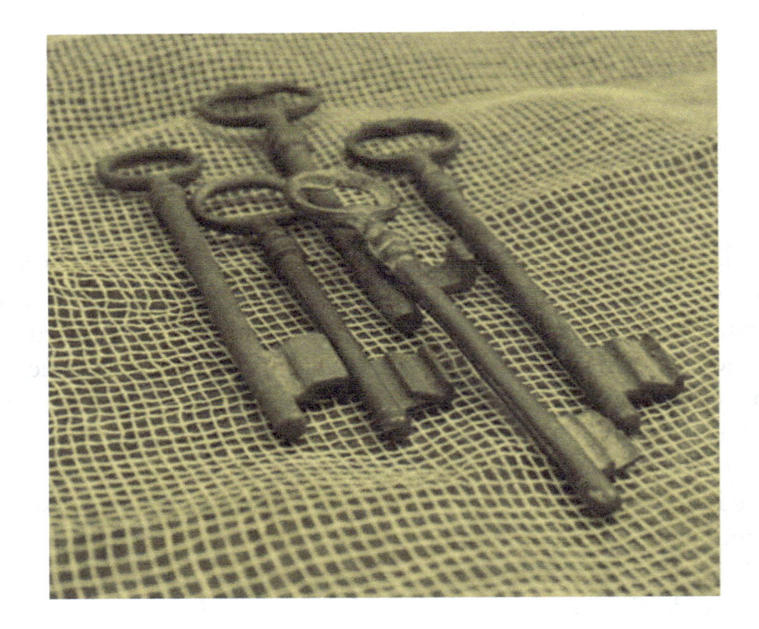

Our main objective is to support students and young professionals by encouraging them to reach out to potential mentors and sponsors as early as feasible to maximize their career success.

We need to clearly define what we mean by mentoring and sponsoring in this book. We provide our own definitions of mentoring and sponsoring, based on our research on these topics and the practical advice shared by the people we interviewed.

Mentoring

Mentoring is a process, in which a person shares her or his insights with another, with the aim of triggering key questions and reflections that facilitate and inform decisions related to career paths and progression. The mentor provides options and highlights hurdles and opportunities, enlightening the mentee in his or her search for advancement, professional growth and—ultimately—success.

Sponsoring

Sponsorship is built on trust and the confidence of the sponsor in the protégé. Often the sponsor feels empathy for the protégé, because of shared experiences or attributes. Typically, the sponsor is a more senior and experienced person. A person acts as a sponsor, when he or she perceives valuable talent and the potential for growth in a junior person and supports and endorses the junior person for new opportunities.

A sponsor may help with introducing contacts, building a network, securing leadership opportunities and obtaining positions of increasing responsibility. When a sponsor provides a recommendation for a person, they risk their reputation by vouching that the person whom they have recommended will perform well.

It is more difficult to find a sponsor than a mentor. Almost everyone from the highest to the lowliest position in an organization can provide useful information to a newcomer as a mentor. Far fewer people are able to support someone for advancement as a sponsor. Before providing a recommendation for someone, the sponsor much trust the other person to perform well.

There are some misconceptions about these two roles. They are not interchangeable. However, the same person may function as both a mentor and a sponsor.

In 2004, David Clutterbuck,[1] a prolific author on topics pertinent to mentoring, who studied mentoring relationships, coined an acronym for what mentors do, based on the word "mentor". That acronym is:

- **M**anage the relationship
- **E**ncourage
- **N**urture
- **T**each
- **O**ffer mutual respect
- **R**espond to the learner's needs.

After analyzing what mentoring entails, we are not in full agreement with this vision, as the "T" for teach and "R" for responding to the learner's need are not necessarily needed in this process. In our experience, the most important elements of a mentoring relationship are encouragement, nurturing, and others not listed in the acronym, including suggesting alternatives and asking questions to trigger reflections in the mentee about the way forward.

We would like to emphasize that career mentoring is not the same as technical coaching. In some organizations, a technical mentor guides a junior person through a structured training program to reach a defined level of proficiency and become "fully qualified" in that occupation. Our focus is on career mentoring including choice of career paths and identification and discussion, but not teaching of how to acquire the required technical skills.

The structured interviews of a wide spectrum of leaders included in our book help in clarifying further the concepts, and at the same time provide a multi-faceted investigation of these valuable career shapers.

Perhaps, a better way to exemplify and explain what mentoring is would be to specify what mentoring is not.

Mentoring is not

- Teaching
- Deciding for the mentee
- Recommending options
- Judging
- Convincing

[1] David Clutterbuck, "Everyone Needs a Mentor: Fostering Talent in Your Organisation", CIPD Publishing, 2004.

- Patronizing
- Friendship
- Psychotherapy.

Mentoring is more about

- Asking questions
- Offering opportunities
- Guiding
- Highlighting options
- Triggering reflections
- Listening
- Humbleness and respect.

Engaging in Mentoring

We have all been in a situation where a person asks for our opinion in relation to a problem. It happens with our family and friends, at professional society events and in the office. Especially at work, once a person builds appreciation or admiration and respect for a role model or a supervisor, it is natural that she or he will look up to this person, and if the right opportunity arises, will ask about an issue of concern, seeking advice.

A less experienced person, who is struggling with a particular problem may think that his or her role model can provide a good solution, because in her or his eyes, that role model possesses a valuable experience and/or has an immense power. They may hope that they can just follow the prescription, without questioning or debate.

Reality is different. A mentor may or may not also be a friend. Most mentors are not psychologists. It is not in the mentor's role to dictate a path, to determine a solution or to decide on behalf of the mentee. That is not how mentoring works and if it happens, it is not good mentoring. A mentor guides but does not make the decisions.

Often younger people and students seek advice from more experienced professionals and/or from their professors. They expect them to explain what to do to solve their problems or guide them towards successfully overcoming any struggles to advance in life and work. If life and careers were limited to a single path, a recipe to follow and mentoring would be a much simpler process, one in which a workflow could be applied with guaranteed success. But life is complicated. Each one of us is unique and our skills, ambitions, gaps and opportunities vary, dependent on context, experience, economic means, culture and business cycles.

The best mentoring occurs when two people, one of whom is usually less experienced regarding a specific topic, explore together an issue that concerns the mentee including the merits of alternative approaches.

Good mentoring enables and empowers self-reflection in the mentee, who, as an outcome of the mentoring session or sessions, comes out stronger and more confident in her or his own preferred decisions about the future. These mentees will come for more in due time, as they will gain an appreciation that the mentors activated something new in themselves stimulating useful insights. Perhaps, the mentors even spurred important and positive insights into their strengths and weaknesses and illuminated an enticing way forward.

The interviews summarized in the following chapters share a wide range of mentoring experiences and outcomes.

Tips for Mentors

The following list of tips intends to help you in your mentoring approach and is intended to trigger your self-reflections about your current mentoring style. Main elements are: Listen, ask clarifying questions, rank and focus, triggering reflections, provide options, and some ideas about how to end the mentoring session.

– **Listen**

As a mentor, you are not supposed to solve the mentee's problems. Even if you know your mentee well and you think you understand their concerns, you need to hear from the mentee, in her or his own words, regarding the topic or issue about which she or he is seeking mentoring.

Open your mind and your heart to understanding what concerns your mentee wants to address. Start with brief greetings and listen to your mentee. She or he may come to you with an endless cascade of words or may be almost mute in front of you. Usually, they will articulate why they want your mentoring. Listen carefully.

– **Ask clarifying questions**

Once you think you have understood the problem about which your mentee is concerned, make sure she or he is aware of the options in her or his power and that she or he has already explored those.

Questions may include, "Have you tried…?" or "Did you contact ….?" Or "what have you done about this?" Make sure you do not re-invent the wheel for your mentee, and that she or he has tried to solve the problem and has explored multiple possible solutions. If they have not yet tried the obvious solution paths within your organization, chances are your mentoring will not be as useful. Also, if you pay close attention, the answers to your clarifying questions will provide an invaluable opportunity for you to explore the thought processes of your mentee and most probably, also her or his preferences, style of communication, level of proficiency, and level of energy.

– **Rank and Focus**

Often a few simple questions will trigger long answers from which you must filter and identify the key issues that would benefit from mentoring.

After you have identified the issue that should be the focal point of mentoring, other complicating issues arise. Rank and select which factor or factors on which you will focus. Remember you only have a limited time for the session.

Prioritize your mentoring to focus on what you think will provide most beneficial outcome for your mentoring session as if you may only have a single opportunity to meet. The mentor should succinctly summarize the issue and get the mentee's agreement on the focus for the session.

You may choose to focus on fundamental problems, such as career preferences or selection of a study path. Other common issues include disagreements with a supervisor.

– **Trigger reflections**

Focus your questions on prompting your mentee to reveal inner preferences, strengths and weaknesses. Common questions include:

- Where do you envision yourself in three (or five or ten) years?
- What would you like to accomplish by then?
- If options and chances of progressing were identical, would you prefer to pursue a technical career or a managerial one?
- Do you envision yourself as an expert in XYZ?

The goal of these questions is to trigger in your mentee the desire to seriously address their long-term objectives, so that she or he understands the inner motivation, drivers and measures of success. What motivates and inspires him or her?

Mired in detail, often we lose track of the big picture. Mentors should ensure the discussion is focused on the big picture and help the mentee to craft a plan to pursue key goals.

When we are fully immersed in a problem, we may be afraid of risks we have already detected. A mentor's questions can help the mentee focus on the main elements and eliminate minor distracting issues. The mentor can provide structure and guide the mentee to address the issues in manageable pieces.

– **Provide Options**

Experience is valuable, but in mentoring, the goal is not about having your mentees repeat what you did, what you lived, but making them aware or different doable alternatives. The mentor may explain advantages and disadvantages of each option, but the mentee is the decision maker.

The mentor supports a decision by suggesting options. The mentee is the one who needs to explore what is best for her or him.

End your mentoring session by summarizing the options discussed, highlighting the strengths and preferences expressed by your mentee and emphasizing the decisions needed.

– **How long should your mentoring sessions be?**

One hour or two hours of mentoring should be enough time to open your exploration path, rank where to focus your mentoring, and provide options

Table 1 Distribution of time in a mentoring session

Time to dedicate to key activities in a mentoring session	
Activity	Time
Listen	5–10 min
Ask clarifying questions	5 min
Rank and focus	5 min
Trigger reflections	15 min
Provide options	15 min
End your mentoring session	5–10 min
Total	50–60 min

for the way forward. The session should be properly closed, with an actionable summary. The mentee should be motivated to explore some of the options identified, to think about them, and to evaluate them in relation to her or his preferences, strengths and gaps. Hopefully, your mentee will find happiness and satisfaction in having a self-decided plan forward for reaching specific career goals.

Even though we are advocates of informal mentoring sessions, and we encourage mini and single-mentoring sessions, we have provided an example of how time can be allocated in a typical mentoring session (see Table **1**).

Differences

Mentoring is a discussion, and mostly benefits the mentee.

Sponsorship is about relationships. It's a bond built on confidence and trust. Sponsoring usually benefits both sides, the sponsor and the sponsored.

Sponsorship depends on the exchange of knowledge and sharing of networks. Because sponsors put their reputations on the line, they need to be confident the sponsoree has the potential to grow and the drive to excel.

Sponsorees do not know what they do not know. They must trust their sponsors to guide them on their professional path.

It's a give and take and is well worth the effort.

Tips for Seeking a Sponsor

Some ways to get noticed and find a sponsor include:

- Ask your mentor to introduce you to key decision makers.
- Join professional networks.
- Express interest in mentoring and leadership development programs.
- Request to be put on high-profile projects or high-visibility teams. This is crucial. Then, it will be up to you to show your value, by performing with excellence or exceeding expectations. Remember, your sponsor has recommended you and expects you to back her or his words with results.
- Use your existing network to get introductions to top-level people. Generally, people are glad to show they are networked with high profile leaders in their profession, and your petition will be welcomed.
- Attend industry and corporate events that draw high-level influencers.
- Volunteer for professional societies, Chambers of Commerce, or community activities that provide an opportunity to work alongside potential

sponsors. You need to liaise with groups that are appealing to powerful people.

- Remember to highlight and communicate about your accomplishments. A sponsor is earned.

The Interviews

Mentoring and sponsoring were the center of our research, and we designed our book to enhance our readers' understanding of both topics. We believe that our book will be of greatest value to early career and mid-career professionals.

While mentoring and sponsoring are valuable at all stages of a person's career, when you are young and have many years ahead, mentors and sponsors can have the greatest impact. More mature people who are considering shifting their course may also benefit as they seek a more satisfying career path that requires opening a different set of doors.

We cherry-picked individuals from many cultures around the world. We included outstanding people in academia, governmental organizations, non-profit groups, sports, and other professions in different sectors and activities.

Table 2 List of interviewed individuals in this book

Country	Interviewed individuals
Armenia	1
Australia	2
Cameroon	1
Canada	1
Italy	1
Kuwait	1
Lebanon	1
Morocco	1
Nigeria	1
Oman	1
South Africa	1
Spain	1
Switzerland	1
Tunisia	1
UK	3
USA	6
Venezuela	5

Fig. 1 Geographical distribution of interviewed individuals in this book

We also spoke with young professionals. A selection of quotes from five of them were included with the goal of providing insights from not only the mentors and sponsors, but also from the mentees. We have explored both sides of mentoring and sponsoring relationships—the givers and the receivers.

The interviews were begun before the global pandemic upset lives and careers worldwide. A few interviews were conducted in person, but most were done remotely while Maria Angela was secluded in Kuwait and Eve in California and the interviewed individuals were working remotely from home.

Our interviews spanned seventeen countries on six continents (see Table 2 and Figure 1). We spoke with twenty seven leaders of global significance and five young professionals. In each of twenty-four chapters, we focus on an interview with one of eleven female and thirteen male leaders.

Each interviewee provided different pieces of the puzzle of mentoring and sponsoring. We were amazed after our decades of personal experience how many new insights we gained from our interviewees. Their wisdom forged the keys that we share with you. We hope the collective wisdom compiled in our book will help you to open the doors to your own success!

Sheikha Intisar Salem AlAli AlSabah

"Sponsoring is to support people to reach their goals."

M. A. Capello and E. Sprunt, *Mentoring and Sponsoring*,
https://doi.org/10.1007/978-3-030-59433-6_2

A Glimpse

Intisar means "triumph" in Arabic and is the name of a Kuwaiti Princess of the reigning family of Kuwait, the Al-Sabah[1] house, who has excelled in humanitarian efforts, Her Highness, Sheikha Intisar Salem AlAli AlSabah.

We are pleased to include in our compilation of insights about mentoring and sponsoring the views of HH Sheikha Intisar, a highly influential leader in Kuwaiti society, with important regional and international initiatives for alleviating the suffering of people afflicted by disasters, conflicts, and wars. She has dedicated her life to advancing humanitarian efforts that seek to provide education, support and equal opportunities for those in need. An extraordinary fundraiser for victims of the Syrian war, HH Sheikha Intisar has been a strong supporter of the International Committee of Red Cross, the Human Rights Watch Kuwait chapter and many other key philanthropic organizations with global outreach.

Mentoring and sponsoring are skills that come naturally to HH Sheikha Intisar.

- Born to the Kuwaiti royal family, Her Highness is the daughter of His Highness Sheikh Salem Al Ali Al Sabah.
- Author of the book, "*Kuwait in 400 years,*" a ground-breaking work that documents the history of her country with 1,300 photos. Also, author of "*Alchemy of Wisdom*", a book about art in the Arab world, which was nominated for the State of Kuwait Award.
- Founder and driving force of the "Intisar Foundation", a UK-based humanitarian organization dedicated to the support through drama therapy of Arab Women who have been affected by war.
- Created and propelling the #1MillionArabWomen campaign with the aim to educate and train 600+ drama therapists and facilitators in 20 years, so they can train women in their cultures in their language and their communities.
- TEDx speaker (2018 and 2012).
- Arab Woman Award 2017 for her prominent part in promoting the role of women in society, the Peace and Tolerance Award, and the Middle East Psychological Association Award.

[1]The House of Al-Sabah is the ruling family of the State of Kuwait. Sheikh Mubarak Al-Sabah (1837–1915) "Mubarak Al Kabeer" or "the Great" was the seventh ruler of Kuwait from May 18, 1896 until his death on November 28, 1915 and is considered to be the founder of modern Kuwait.

- Held the exhibition "Women in War" in Kuwait in 2017 with the International Committee of the Red Cross, highlighting the challenges of women in war-torn areas.
- Member of the Executive Committee of BACCH, the first paliative care hospice in the Middle East, and an executive member of KACCH, the first non-governmental organization in the Middle East that takes care of children in hospitals.
- Member of the Board of Trustees of the Lebanese American University and the Faculty of Social Sciences, Kuwait University.
- Founder of Lulua Publishing, an editorial house focused on publications that stimulate, inspire and promote psychological and physical well-being.
- Majored in political science at the Faculty of Social Sciences at Kuwait University.

A Personal Snapshot

This interview provided a glimpse into the insights and perspectives of a woman with genuine passion to help others.

She has left her comfort zone to pursue noble causes in complex political settings and is truly inspirational.

The Interview

Q. Tell us about your own mentoring and sponsoring experiences.

I think everyone is mentored throughout their lives. I ask myself, if I have really been mentored, and I think it has always been unconscious. All I had to do was to keep my eyes open, pay attention to how the experienced people were or are working, ask questions, and that was automatic. Mentoring doesn't have to be conscious.

I was chosen to be a mentor in a formal program and was overwhelmed when I realized we were all required to follow a structure in a very formatted context to be mentors or mentees. Actual mentoring for me is something else.

Our mothers mentored us, our family, professors ...they all mentored us. Mentoring Is constant learning from people and especially from people you admire. Mentoring is an ongoing process.

People think I am a good mentor, but I don't know why. I didn't learn. I don't consciously know how I manage people. I just do. It is not a monthly shift, or a scheduled activity. If it is too structured, mentoring loses its essence.

It is like training, nowadays. Training is not a onetime encounter or activity. You do training in small bits because you need time to digest.

Mentoring should not be structured. I cannot imagine a situation where you just sit with someone for an hour and can take away all the learnings. No, it cannot happen like that!

Instead, I believe mentoring is something for which you search. In the USA, for example, people will accept a post as a personal assistant with a very small salary, just to be shadowing a CEO and learning from her or him. They get to see what leaders do firsthand.

Mentoring is definitively not an hour meeting. It is a more like a long-term relationship, and it can be as simple as accompanying someone you admire. That would be a mentoring lesson. See and learn.

It would be brilliant if you can learn in one hour and then another hour the following week. Once a month is not the way to go, in the sense that I doubt you can learn something useful in that hour from a mentor. It could work, but it is difficult. Why? Because a structured approach is not human. Things do not work when you take away the human factor. It becomes more like preaching, not mentoring.

Mentoring is not in the mind. It is only in the heart. It must go to the heart.

For example: I know exercise is good for me. I want to go and do my exercising, but I don't. This is a typical example. You understand it, but you don't feel it.

It is a determination that is only a mental determination. It is not deep inside in your heart. So, I just don't go to the gym, because I have not made exercising a priority.

Another example. I know people who read hundreds of books, but they still are in their comfort zone. They are the same people as before, from ages ago. They are not going to change, because they are not invested in what they are reading. The books have not influenced them.

***Q.** Have you mentored? How do you mentor?*

I know I do it, because people tell me I have triggered in them step changes. It is not a one off for me, so this happens with people who are close to me for extended periods of time.

Unless mentees are open to observe, they will not learn.

Q. *What do you think about mentoring online?*

For sure! I use my social media as a platform to reflect on what I would like people to do. Like a role model, because I have been told and consider myself to be a role model. Because of this, I want to be a good mentor, and in my posts, I don't reflect negativity or complain, especially in my Instagram posts. Never! Although sometimes I may feel negative, I want them to take positive messages from me.

Nevertheless, I don't think of myself as a mentor on-line. When people are commenting that they are inspired, I think "*Oh, wow! why? I am just sharing my life*".

Today, I got a message via Instagram, "*Please continue doing videos. They inspire us.*" These kinds of messages are a constant, and I do not want to let them down. I must continue doing what I am doing.

Q. *Please tell us about your sponsors.*

I have several sponsors. In the USA, I have a sponsor in the Citibank, and she is introducing me to key liaisons who will guide me to fulfill my aims with my Foundation.

I have another friend, who is sponsoring me specifically in the UK.

Sponsorship in the end is when you find someone who opens doors. I have my key, but I don't have all the keys needed for all the doors.

I have the keys to my doors in Kuwait and I can open doors in my country. But I don't have the keys for doors in other countries, and my sponsors open doors for me in other countries. Each one of us have our own keys.

Sponsoring is supporting people to reach their goals. You have this desire to support them because of what they do. Whatever it is. The sponsor is invested in the sponsoree.

Q. *Is there any anecdote you would like to share?*

Life is so huge. You don't expect things to happen, but then, they are there.

The Deputy Head of the International Committee Red Cross (ICRC) in Kuwait, approached me and they invited me to attend in Dubai a major event to celebrate a partnership with Philips. I asked her "*Why do you want me to attend*?" I was hesitant, as I was not part of the circle of international organizations yet. This was about five years ago, and at the time I was not engaged in the humanitarian world.

She replied, "*You are the perfect example, you are from a Royal Family, you are active, you are a woman, and we want you to be there.*" So, as she was so adamant about this, I went. The event was under the Patronage of Princess Haya, who was at the time the wife of the ruler of Dubai. But at last minute, we received the news she could not attend, and I got to be seated where she was to be seated. So, I was beside the VicePresident of ICRC, Madame Christine Beerli. We got along so very well; afterwards, every time they had an event, she would call me and ask for me to attend. That is sponsorship.

She liked me. She thought it was beneficial for me to attend the ICRC gala events. They wanted to do a round table in Kuwait and I was appointed to organize it.

Q. *What do you think about this kind of sponsorship?*

This kind of sponsorship was not formatted or scheduled.

So, I did the IRC (International Committee of the Red Cross) in Kuwait. We had 12 very influential, humanitarian, socially influential women attending from Kuwait. From Switzerland, we had Madame Beerli, who flew from Geneva and her deputy, who was based in Syria. In total, there were four people from ICRC and the rest were a dozen Kuwaiti woman leaders.

We came up with recommendations for what we could do for women and children and an art exhibition about women in war. This is what gave me the idea to support women psychologically. This is how my foundation was born.

Q. *Do you take a risk when you are sponsoring someone?*

Well, I think it is important to keep sponsoring and I do take risks. Sometimes, it will go fine, and others not, but I still make it. I still sponsor many people, even if there are times when I discover that they are not up to it. I learn from these experiences. I go back, I take full responsibility and try again.

There is a saying in Arabic that goes, "*You do not know someone until you travel with him.*" I like it very much, because it is true. When you travel with a person all day long, you get to know the good and bad aspects of that person. How they treat others. How they think.

Life is about taking risks and making mistakes. That's the life of a doer. So If I introduce my sponsoree to a friend who I think would benefit from this introduction, it might go amazingly well and my friend would thank me for introducing them to my sponsored person. Or it might go wrong, and I will take responsibility and apologize to my friend and find ways to make amendments But thinking of the opportunities I am creating by making these connections outweighs the mistakes. I have learned to judge better and so my mistakes are way less.

If I sponsor someone, I am taking a risk. What I am doing is that I am supporting even with risks. I may make a mistake, but I will keep sponsoring anyhow.

Q. *Do you think we tend to sponsor only those who are like us?*

It depends. I know many boards of directors and in some, yes, all members look alike. Twins all over! But the boards to which I belong are very diverse. The Board of Trustees of the Lebanese American University (LAU) in Lebanon is truly incorporating a variety of nationalities and backgrounds. I think that integrates and produces better decisions.

When I hire people for my initiatives, I aim for people who are not like me, but who could complement me and people who are better than me. This pushes me to be the best I can be. I don't hire people equal to me.

I think that boards where all member look alike are a result of the insecurity of the leader.

Just yesterday, I attended a management meeting of my Unit Heads of my foundation. One of the managers was adamant. I needed to express I was not in agreement, but she said something like, "*I like to fight back, because I want to do the sales and I need X, Y, Z elements to achieve that.*" I found myself liking that kind of challenge and push back to my ideas. People must fight for what they want and think. When the leader is weak, they hire people like them, so they do not face challenges or different perspectives.

At the Board of LAU, we are not alike in any way and we get along well, because the chair is amazing. Before, there was a lot of politics in LAU, but the new chair brought more people who are amazing. The group is very dynamic. The stronger the leader, the more diverse the board.

Q. *How do you handle your role modeling with so many groups of people?*

(Sheikha Intisar stops for a moment, thinking)

I never thought of myself as a role model, but I now know that I am, because lots of people stop me either in the street or send me messages on my social media, to tell me how I inspire them.

What I realized is that just by being me, I become a role model. It is not through acting. If I was acting, most people will feel it, and I would be an impostor. But eventually all is always revealed and just by consistently being me, I allow others to be themselves and enjoy the feeling of liberty and authenticity this entails.

This is the best gift one can give to oneself and to others. The role modeling of being themselves, and even if I have many hats, I am still me in different situations, Just with a different flavor added.

When I go to schools and I find the kids, I hug them. I am very spontaneous. I usually tell them, "*Go after what you want.*" Sometimes they come and hug me.

It is very interesting. Even the most extroverted person can be shy.

When I find someone who is shy, I tell them that mentally, you must "go" to a place that feels comfortable to you.

Q. *Do you remember any special sponsor or mentor of yours?*

I remember in high school I was not so good in physics. It was the only subject in which I was bad. What the school did was to shape a group with all the not so high graded students, and they gave us the best teacher ever. I went from Grade F to C within one year. I didn't really study. If I did, I would have gotten an A. The professor was so good! That teacher was special for me. His name was Mr. Boyd.

Another teacher important for me as a mentor was my Arabic teacher. I became the first in the class, just because I liked this teacher so much.

After these experiences, I had a good grade point average (GPA), and could study at Kuwait University, and I chose Political Sciences.

Q. *Is gender or age important in mentoring? And in Sponsoring? Females mentoring females, or more-experienced professionals mentoring young?*

Not at all. As an example, one of my best sponsors is a man, the head of the Kuwait Planning Ministry, Dr. Khaled Mahmeed. He sponsors me and our program.

Q. *What differences do you find in mentoring or sponsoring between Kuwait and the rest of the world?*

It is simple. Kuwait is smaller, and perhaps with only one degree of separation, so everyone sort of knows everyone else, this makes finding an ideal sponsor easier and getting to them simple.

Q. *Why we don't have more women in leading roles in Kuwait?*

This is not a phenomenon only pertinent to Kuwait. This is a global situation. Our priorities are different than men's priorities. Our priorities are basically seeking to balance career and family.

I was reading the other day about the coronavirus pandemic in China, where about 70% of health workers are women, and there were family issues with long working hours, because they were leaving their children at home with no care or support. This is hard psychologically for the women as well as for the children.

I think is important to include into the understanding of women's leadership the necessary gap for women's maternity which is about ten years. Women in these ten years have a family. In Kuwait, this is generally from 21 to 34 years of age.

Q. *Do you have children?*

Yes, I have four daughters, and the youngest are pursuing their BA in the UK.

Q. *What do you wish you would have done differently in reference to mentoring or sponsoring?*

I would have observed and asked questions earlier instead of thinking, "I know things." With time, I have realized I don't know things and even if I do, it is better for me to keep quiet and listen. Then, I always learn.

Q. *Are you transferring this to your daughters?*

As much as a mother can.

Q. *And about sponsoring, would you change anything?*

I would listen to myself more. I would commit to listen more.

Q. *Did you enjoy any mentoring or sponsoring experience? Tell us more about those joyful occasions!*

All my life I have been sponsoring or sponsored, mentor or mentee.

I have been mentored by my family. I learned from my teachers and then friends. I was also mentored by my aunts and uncles when I see how they interact with people.

I have been mentored by all the people I ever worked with, just seeing how people deal with things and events is a first hand mentoring experience.

I have also been sponsored by lots of people who saw something in me and wanted to grow it.

I learn from everything and everyone. For example, I have a gardener and I just look at him when he is doing the gardening and how he is really in the moment and enjoying his work. As soon I see him, I feel his joy for having a job and how he enjoys it and how he works with a smile on his lips, and how grateful he is for everything, big or small and I learn from him. That is a sort of mentoring. For sure!

In short, I learn from everyone. Children can be a treasure trove of learning. They enjoy single moments. Corona virus will change priorities for everyone, for countries, too. It is going to be a painful transition.

A Shared Selfie

- **Your favorite role model**: I can't think of one person. People are my role models. Everyone. I learn from all, as I see in every single person the good and the bad parts of me.
- **One word to define your experience with mentoring**: Sharing.
- **One word to define your experience with sponsoring**: Supporting.
- **Who would you have liked to ask more sponsoring from, but never asked?** I would like to have asked for more sponsoring and more mentoring from my own family. I put very high expectations on my family, and they also put those on me.
- **Who would be an ideal mentor for you in this moment of your life?** Sheikh Mubarak, the founder of Kuwait. He was so smart and so brave, and we all have a huge admiration for him in Kuwait. He is one of the greatest icons of our country.
- **What would you like to ask him**? I would ask him how he could handle so many fronts of diplomacy, political actions and even war at the same time.
- **Do you think he could have imagined the grandeur of today's Kuwait?** During his time, he made Kuwait grow five times its size. He had great aspirations. So, I suspect that yes, he would have envisioned a great Kuwait.
- **What do you think he would ask you?** I would have to think about it, but most probably, he would ask me if I have thought about expanding further what I can do at the global scale.

Afterthought

We left the interview with birds' singing. Yes, there are many birds in springtime in Kuwait. It was a beautiful garden and helped us to understand the joy of life HH Sheikha Intisar was referencing. We need to learn from everyone around us and make sure we have the right keys to open the doors of opportunity.

We really liked the analogy of keys and doors HH Sheikha Intisar provided and realized it is widely applicable.

Looking at our set of keys and our own hearts, which keys are we missing?

Dr. Rita Rossi Colwell

"In sponsoring you balance prudence and risk"

M. A. Capello and E. Sprunt, *Mentoring and Sponsoring*,
https://doi.org/10.1007/978-3-030-59433-6_3

A Glimpse

At Latitude 78°2′ S and Longitude 161°33′ E in Antarctica, where average annual temperatures are −20 C, there is a rugged rock massif, about 4 nautical miles (7 km) long, that rises to 2,635 m, in the area called Victoria Land. It is the Colwell Massif, a geologic site which was named after Dr. Rita R. Colwell, in recognition of her work in the Polar Regions.

Rita Rossi Colwell is an American marine molecular bacteriologist, who received the 2006 National Medal of Science of the United States of America. She was the first female director of the USA National Science Foundation from 1998 to 2004. Her contributions to the research of global infectious diseases, water, and health include the development of an international network to address emerging infectious diseases and water issues, including safe drinking water for both the developed and developing world.

We were thrilled to interview this extraordinary scientist, whose interests include K-12 science and mathematics education, graduate science and engineering education, and increased participation of women and minorities in science and engineering.

- Distinguished University Professor at the University of Maryland at College Park and at Johns Hopkins University Bloomberg School of Public Health, senior advisor and chairperson emeritus at Canon US Life Sciences, Inc., and president and chairperson of CosmosID, Inc.
- 2006 National Medal of Science USA presented by President George W. Bush.
- The eleventh and first female director of the USA National Science Foundation (NSF) from 1998 to 2004.
- Among the honors she has received are the 2006 "*Order of the Rising Sun, Gold and Silver Star*", of the Japan Society for Promotion of Science; the 2010 Stockholm Water Prize Laureate; the 2016 Malaysian Academy of Science "*Mahthir Science Award*" for Scientific Excellence in Work in the Tropical Regions; and the 2016 "*Prince Sultan Bin Abdulaziz International Prize for Water*", Creativity Prize, Saudi Arabia. She was knighted in 2017 as *Chevalier de la Legion d'Honneur*, France.
- She is Chair of the Research Board for the Gulf of Mexico Research Initiative through 2020, and served as President, American Society for Microbiology (ASM), President of the American Association for the Advancement of Science (AAAS), President, American Institute of Biological Sciences (AIBS), President, American Institute of Biological Sciences (AIBS).

- She has held multiple advisory positions in the U.S. government, nonprofit science policy organizations, and private foundations, as well as in the international scientific research community.
- Author or co-author of 19 books and more than 800 scientific publications.
- Colwell has been awarded 62 honorary degrees from institutions of higher education, including her alma mater, Purdue University.
- She received her Ph.D. from the University of Washington in Oceanography.

A Personal Snapshot

Dr. Rita Rossi Colwell is a woman of many "firsts," with many awards and recognitions that highlight not only her accomplishments as a scientist but position her as an inspiration for the younger generations of women rising the scientific ladder in their organizations. Younger women benefit from the battles won by those who came before them. She shared a wonderful history of mentors and sponsors to whom she expressed gratitude.

Interviewing Dr. Colwell was a delight for us, because as women in STEM careers and life-long promoters of STEM careers, we were aware she is one of the top role models for young women studying sciences, specifically biology.

The Interview

Rita was very approachable and open. Given her importance and prestige, she showed true leaders are indeed accessible, supportive and welcoming.

Q. *Rita, please tell us about your own mentoring and sponsoring experiences. Did any particular mentor or sponsor play a crucial role in your career?*

I'm the daughter of Italian immigrants, who came to the US. I was born in the US. My father was very, very supportive of women and education. That made a huge difference. He insisted I get as good an education as possible and was always encouraging.

My mother died when I was 14. My aunt, who helped to care for my siblings and me, tried to convince my father to send me to secretarial or nursing school, because those were the most common careers for women at that time, but my father stood firm that I should go to college.

Therefore, my dad was my first important sponsor in my professional career.

I had a series of teachers, and some were truly extraordinary. As a sixth grader, I had to take a test that I presume was an IQ test, and all I know is that afterwards, the principal of the school took me aside and was adamant that I had to go to college. That was early on when I was probably only 12 or 13 years old. And that was perfectly fine with me!

Later, in high school, I also had some excellent teachers, some of whom were supportive of women. I must admit that only the female teachers were supportive of women, not the male teachers. My biology teacher, for example, was a man, and was also the coach for the football team, so he was more interested in football and supported male classmates.

I was very lucky as an undergraduate to have an advisor, Dr. Alan Burdick. He was a remarkably kind and smart man and a fine mentor for me.

When I was finishing college, I was accepted to medical school, but in April of my senior year I went out on a date with the man I married two months later. Jack, my husband of 62 wonderful years of marriage, passed away two years ago. We had two daughters. So, there I was, married before I finished college. We decided it would be a good idea if we stayed at Purdue and both do a masters and then, pick another university for my studies, where he could continue in Physics.

I asked for a fellowship and was told they wouldn't waste fellowships on women. So, I went to my undergraduate advisor, and he said I could do a masters with him on genetics, using fruit flies for lab experiments. This research was in the early years of genetics, when molecular genetics was being launched.

Jack applied to University of Washington in Physics, and I applied and was accepted into medical school, but to register in that medical school I had to be a legal resident of the northwest of the USA, and I wasn't. So, I went to University of Washington, because my former undergrad advisor knew Dr. Hershel Roman there, who was a famous professor.

I got caught in a fight between two professors. It was obvious the dilemma would not be resolved, and I wouldn't be able to work with the person my undergraduate advisor recommended. So, I ended up with Dr. John Liston from Scotland, who had just arrived at the University of Washington to set up a program in marine microbiology. He became my mentor.

I could not establish residency in time to get into medical school, so Dr. John Liston helped me earn my Ph.D. and later was instrumental in my obtaining a position as a postdoc in Ottawa, Canada, where my husband was offered his postdoc. There was another huge problem, because even if I was accepted for a postdoc, I was told that due to nepotism rules I was ineligible, so my advisor, Liston, solved the crisis by enabling me to get a grant from the National Science Foundation and go on leave from the University of Washington.

Dr. Liston turned out to be a lifelong mentor, who made many things possible for me.

During my career I had good friends who made it possible for me to succeed when things were going wrong. I was lucky.

After the postdoc, my husband got a job at what is now the National Institute of Standards and Technology, so, I had to get a job in Washington. A friend introduced me to Dr. George Chapman, who was starting a new department. We met in the lobby of a hotel during a conference, and he offered me a job on the spot, at Georgetown University. That is how my career as an academician began.

Later, I moved my laboratory to the University of Maryland, where several years later I was appointed Vice President for Academic Affairs and Provost. A few years later, I was nominated by President Clinton to be Director at the National Science Foundation (NSF). I returned to academia after my tenure as Director of the NSF with a joint appointment at Johns Hopkins University and the University of Maryland.

Q. How did you get President Clinton to appoint you?

I had served on many advisory committees at the National Science Foundation. I had also served on the National Science Board and I had been very active and founded the Biotechnology Institute at the University of Maryland. I became active working with state legislators due to my work, and continued my research on water and epidemics and through that work made friends with representatives and senators in Congress.

Before being appointed Director of NSF, I had previously been asked if I would consider an appointment as deputy director of the NSF, and I very politely said no.

Q. Why did you decline?

I realized that if you want to make a difference you have to be in charge. I explained my reasons and said that if the opportunity to be director came up, I would be very interested and pleased to accept.

Dr. Neal Francis Lane[1] was the NSF Director at the time, but he was about to be appointed Science Advisor to the President, and therefore a Director for

[1]Dr. Neal Francis Lane is a U.S. physicist and Senior Fellow in Science and Technology Policy at Rice University's Baker Institute for Public Policy and Malcolm Gillis University Professor Emeritus

the NSF was needed immediately. Thus, I received a call from Vice-President Gore in December 1997, and I accepted the invitation to serve as Director of NSF. My appointment was announced in January 1998.

Q. *Please tell us about how the directorship was offered to you?*

This is probably one of my best anecdotes.

> I was entering my office at the university and my assistant said, "*The Vice-President is on the phone*".
>
> I wanted a clarification and asked, "*The Vice-President of what?*".
>
> My assistant replied, "The Vice-President of the United States of America."

So, I got on the phone, and it was Vice President Gore's secretary, who kindly informed me that if that was not a good moment to talk, they could call me back. Of course, I confirmed it was a good moment!

Q. *Now, please tell us about your own process for sponsoring?*

I have been a sponsor and have played a role in several appointments of women to chair positions and other appointments when I served on search committees.

I have always tried to be low profile but provide the nudge to get the best qualified person into the position. I try to do it so that I am behind the scenes. It is better for women helping other women to not do this very openly, because men do not look on it favorably when women are agressive.

Therefore, I prefer to be quiet and subtle when I sponsor someone.

Q. *Have you sponsored numerous women?*

I have served as an advisor and graduated over 60 Ph.D.'s, of which almost half are women, so I think I've been successful.

Q. *Do you want to share any cases of your mentoring of women that you recall for any specific reason?*

Thank you for this question. One of the things I find about women students is that they are always very hesitant and often do not appreciate their own talent.

I have had students who thought they were not smart enough to pursue a Ph.D. I had to literally push them, to convince them they were perfect for it.

---

of Physics and Astronomy Emeritus at Rice University. He was Science Advisor to the President during the Bill Clinton Administration (1998–2001).

Now, it is very rewarding to see that three of them have to the elected to the National Academy of Sciences.

Q. How do you recognize talent?

Some women are very shy, and some are willing to speak up, so you cannot take communication skills as the only hint or characteristic. It is the competency and ability to work hard and to be curious that you notice in an individual. Also, you can detect who wants to learn and is not superficial in their research and goes into great depth. It becomes obvious who the real winners are, at least to me.

Q. Your support of women is indeed powerful. On other topics, more within your personal sphere, have you been as good sponsor of your daughters as your father was with you?

I have two daughters and one is a medical doctor, who runs a clinic. My other daughter is a botanist at UC Davis – she is an expert on California wildflowers.

Q. Do you think that sponsoring someone is taking a risk?

I think it is a mixture of prudence and being accurate and wise in the risks that you take when you endorse or sponsor someone. You cannot go through life without taking some risks.

There are some risks that you take and some that you don't. Experience is having the wisdom to understand the difference.

One of the issues is convincing someone whom you think has tremendous promise that they have promise. It is more of an issue for women who have a potential for an intellectually rich career. Eventually, they get the message.

Q. Do you think mentoring has to be in person, or are on-line, digital mentoring solutions equally effective?

The COVID19 pandemic made all of us embrace the online solutions. In the last three months, I have been mentoring online. It is better than nothing. But it is definitively not my favorite way. I really enjoy my classes, helping my students in the lab, looking over their shoulders, asking them questions.

Q. Did you encounter hurdles or barriers to finding mentoring or sponsoring?

From the beginning I was very determined to be successful in the things I wanted to do.

Q. *Were you resilient when you encountered adversity?*

Resilient? Yes, when I get angry, that empowers me. I have learned how to channel my anger into constructive response.

Academia is now different. Society is different. We have seen a cataclysmic change towards minorities. There is acceptance. The world has changed so much, and we humans need to change with it. Some humans change more rapidly than others.

Q. *Do you think the #MeToo movement helped or hampered the inclusion of women in the workforce and in society?*

We must be very careful of slogans. I think that slogans raise antagonism. The slogans you use may become boomerangs and be used back at you as a weapon.

Q. *What questions should Mentees ask you?*

What kind of tools are needed for success? What programs are there to elevate my competencies? These types of questions would be beneficial for a mentee to envision paths ahead for his or her career.

Q. *Is gender or age important in mentoring? And in Sponsoring? Females mentoring females, or more-experienced professionals mentoring younger colleagues?*

One of my students was older than myself. I was her mentor, and we did not have any issues. She was excelent and was a co-author of a major book on proteins. She came to my lab wanting to do a Ph.D., so it was no problem at all.

I do not have issues with any form of gender. I embrace and support the nominations for honorary degrees of minorities including LGBT. We need all the talent available!

Q. *Who was a special mentor for you?*

Dr. John Liston. I could not have been successful without him. He was loyal and supportive and gave me all kinds of opportunities. He found ways around the barriers.

Q. *What do you wish you would have done differently about mentoring or sponsoring, as a mentor or as a mentee?*

I would have been more understanding and patient. I tend to be impatient, because I like to work quickly and get things done. I wish I had taken time to think through some of the situations I have faced in life.

I would have savored the moment much more.

Q. Was your husband a mentor for you?

He was wonderful. I was extraordinarily fortunate. He was tall, handsome, a sportsman, clever, loved our daughters. A good physicist and interested in many things rather than in single topics, so, very flexible. He was a champion golfer, almost professional. Then took up sailing, as it was a way to be with the family all together. A family man. I miss him very much, and yes, he was a great companion and mentor.

A Shared Selfie

- **Your favorite role model**: There are many people I love and respect. Perhaps Marilyn T. Miller, who was my roommate during our college years, could be considered my role model. She went on to medical school and became a pediatric ophthalmologist. We stayed best friends, throughout our lives, travelling abroad and developing our careers in similar ways. She won the National Academy of Ophthalmology's top award, and many others, from the medical association.
- **One word to define your experience with mentoring:** Enjoyable.
- **One word to define your experience with sponsoring:** Rewarding.
- **Who you wanted to be your mentor/sponsor, but you never had the chance to ask?** I always asked. For example, when I wanted the university president to mentor me, he appointed me as his VP and provost.
- **Who would be an ideal mentor for you in this moment of your life?** Eleanor Roosevelt
- **What question would you ask this mentor?** I would ask her how did she put up with so much?
- **What question would you like Eleanor to ask you?** How are you so lucky that you married so well?

Afterthought

We realized the interview with Rita R. Colwell let us peek at the life of a very fortunate, happy, and grateful person. We were fascinated by her acute insights about the role of women in society, the love she expressed for her late husband and her two daughters, and her obvious openness to changes in society.

We found her to be a very open and generous person, who maximizes the value of her position as a role model, because of the many awards she has received and important roles she has held in science. She is an outstanding example of a pioneer and trailblazer woman, who shines with her own bright light.

It was a privilege to interview her and include her insights on mentoring and sponsoring in our compilation. We saved this anecdote for last, because this reflects the Rita we discovered: a spontaneous scientist, who enjoys the big and small moments of her life:

You have so many anecdotes! Are there any others you would like to tell us about?

It would have to be one involving my husband. He came to the lab where I was responsible for the stock collection of fruit flies, *Drosophila melanogaster*, used for genetic research. My job was to feed the fruit flies a corn-meal honey puree daily. Fruit flies carry mites, which can make you itch if you don't handle them carefully.

I had about 200 different genetic stocks of flies in the collection. Jack, my husband, entered the lab and, without gloves, he picked up a bottle of the flies to inspect them, while he waited for me to finish up so we could go to dinner. After a few minutes, the skin of his hands and arms began itching from the mites. It was memorable enough for him that he vowed he would never again visit me in that lab…and that he had another reason to prefer to study physics!

George Steinmetz

"Mentors showed me what I did not know".

M. A. Capello and E. Sprunt, *Mentoring and Sponsoring*,
https://doi.org/10.1007/978-3-030-59433-6_4

A Glimpse

A single glance at certain photographs can forever alter our understanding of important events, technologies, landscapes, cultures and more. The people behind those life-changing images are highly skilled photographers with a special gift and ability to capture an angle, a moment, a scale that can explain an idea, a fact, or an issue with their camera.

George Steinmetz is one of those gifted photographers and has one of the most striking and meaningful portfolios of aerial photos. He has captured unforgettable images of a wide range of subjects including oil exploration, secret places in the Sahara and Antarctic, the largest coal mine in China and the wonders of robotics.

Steinmetz has more than 1.1 million followers on Instagram. That number grows every day, as his images and commentaries raise eyebrows and attract fans globally. One of his pictures is considered to be one of the best photos ever in National Geographic. It is going viral with millions of views. Chances are you have already seen this photo in your own social media channels: It is a caravan of camels, which at first seem to be black camels walking in a sand desert. Then you realize those are very long shadows of white camels at sunset. The photo was taken by Steinmetz from his motorized one-seat paraglider.

Since his first assignment for National Geographic in 1987, he has been an important contributor to the National Geographic Explorer TV channel, and created more than 40 major photo essays for National Geographic including three cover stories and 25 stories for GEO magazine in Germany. Steinmetz won first prizes in science and technology in 1995 and 1998 from World Press Photo and other honors including Pictures of the Year, Overseas Press Club, and the Alfred Eisenstaedt Awards. He is the author of two books and an avid documenter of climate change and global food supply.

- Born with a free spirit in Beverly Hills, California
- Published his photographs in The New Yorker, Smithsonian, TIME, The New York Times Magazine, and is a regular contributor to National Geographic
- His posts @GeoSteinmetz and @FeedthePlanet in Instagram have 1.1 M and 200 K followers as of May 2020.
- More than 40 major photo essays for National Geographic including three cover stories and 25 stories for GEO magazine in Germany. Contributor to New York Times Magazine.

- Two first prizes in science and technology in 1995 and 1998 from World Press Photo.
- "The One Club Gold Cube" Award, for his work on large scale food production, in addition to Pictures of the Year, Overseas Press Club, and the Alfred Eisenstaedt Awards, the 74th Annual POYi Environmental Vision Award; Overseas Press Club and Life magazine's Alfred Eisenstadt Awards.
- Has explored subjects ranging from the remotest stretches of Arabia's Empty Quarter to the largest coal mine in China.
- His work has been exhibited in Dubai, the Brookfield Winter Garden in New York, the Arizona-Sonora Desert Museum, the Konica-Minolta Plaza in Tokyo, as well as public venues in Houston Denver, Los Angeles, Toronto, Stuttgart, Expo 2015 in Milano, the Triennale di Milano and twice in the Festival Photo La Gacilly in France.
- Was featured in TED Global Tanzania where he presented his work on Africa.
- He is the author of five books; "African Air," "Empty Quarter," "Desert Air,", "New York Air", and "The Human Planet."
- Currently focused on remote landscapes, our changing climate, and how we can meet the ever-expanding food needs of humanity.
- Graduated from Stanford University with a degree in Geophysics.

A Personal Snapshot

In the interview chapters, we include "a personal snapshot", to summarize our personal impressions and previous knowledge of the interviewed person. Maria Angela has been following George Steinmetz for several years in her Instagram account because of her dual passions about the environment and the scale of things. Maria Angela's husband is a geophysicist who is also a photography aficionado. This has motivated her desire to understand what it means in artistic photos to have "a photographer eye".

Eve and Maria Angela were both surprised to discover George Steinmetz was a geophysicist, as they both are! And not only that, but Eve was a graduate student at Stanford while George Steinmetz was an undergraduate there.

Although photographic and geophysics careers are very different both need mentoring and benefit from sponsoring. Steinmetz's insights were refreshing, and the interview was a fascinating succession of anecdotes and stories of his field work high in the skies where the he took photographs which magically provide to millions of people new appreciation of the world around us.

The Interview

Q. Please tell us about your own mentoring and sponsoring experiences.

I started my career in photography the year I dropped out of Stanford and went hitchhiking around Africa. I did not have much money, but lots of enthusiasm and a camera. During that restless journey I taught myself about photography and had a great experience, but when I returned, I could not make a living from photography.

I realized I had to seek out more knowledgeable people to learn the profession and to swallow my pride to ask what I did not know. I started out working as an assistant for a a studio photographer in San Francisco, California. It was not a dream job. I had to take out the trash, buy lunch and stuff like that. The studio for which I worked was doing what was called "still life photography," a fancy term to say they photographed objects sitting on a table, including silverware, books, and other objects. They would position a desktop computer on the table, because it fashionable at the time.

I did not want to photograph computers on a tabletop, so I sought work with a "location photographer," the magazine photographer, Ed Kashi, who photographed the young titans of Silicon Valley and technology including Steve Jobs and Bill Gates. I was literally his baggage carrier. At that time, photographers had to carry a lot of heavy equipment, including not-so-portable lights, backdrops, heavy lenses, and so on.

We became good friends, but he fired me after three months because I would not follow directions. Luckily, I maintained his friendship and he passed jobs on to me. They were the unglamorous jobs he did not want and gave to me as hand-me-downs. I started to work my way up the food chain of photography and Ed showed me what I did not know. That experience was key.

My advice to people is not to go to photo school, but to go to the field with someone who knows the trade. You must improvise.

For example. you go to a boring place with a grim boring subject with boring clothes, boring everything and he only gives you only fifteen minutes to take a picture for your story. You must make something out of nothing.

The typical question you ask yourself is "*What am I going do now?*"

The big shot subjects were available for only ten to fifteen minutes, you must be creative, make them look different or important. It was an incredible learning experience. It was a difficult sink-or-swim time. Really tough!!!

Q. How did you solve the gaps?

Sometimes, for these hand-me-down sessions, I would call Ed for advice. I would do some preparation and a test shoot beforehand. You must prepare and try to figure it out. Then, you must improvise on the spot. Creativity is important.

I would like to share with you an example of what creativity is. I was photographing a fashion show backstage in Milano, with Gianfraco Ferré, one of the masters of fashion. A model passed by, perfectly dressed in an evening gown. The stylist stopped her seconds before she went on the runway, and Ferré asked an assistant, "*Pass me an organza scarf!*"

Then, Gianfranco himself shaped this organza piece on the model, and the whole thing transformed into an artwork. That is creativity in action! Fashion geniuses do just that. Or stand-up comedians-. On the spot art, on the spot creativity.

It takes the right talent to have the creative intuition of what works.

We need to dig more into the importance of creativity and the ability to pull something together out of few things to enhance a profession, a product. Something must be said for talent and improvisation. You see that in geniuses in all walks of life. You must be able to do something along those lines to be successful. Talent and creativity have a close relationship with improvisation.

I cannot tell you how many times I took off in a helicopter with one idea or objective in mind, but things happen, and you must find a way to substitute, to improvise, to make it work.

Q. Your work is highly specialized in environmental and supply chain for food. How did you find your niche?

It was an evolution. You go out and look, follow your passion. One thing leads to another. I got interested in food after flying over deserts. Visualizing landscapes in a different way makes you think about how to feed humanity in the future.

Then, National Geographic asked me to work on a project specifically about that. They saw I could visualize landscapes in different ways. An expert explained to me that because of population growth, we need to double the food supply. I realized I needed to look at large scale food systems. Mega farms. Mega scales.

If you photograph two zebras, that is boring. If you photograph 10,000 zebras, that is a picture! So, I wanted to photograph mega scales, I needed to photograph how 150,000 cows were being fattened for our food supply. I found very large farm systems of this magnitude and began documenting everything needed for those systems to work including their own food supply and all the related waste. In the short run, these operations may be profitable and sustainable, but not in the long run. Mega foods systems are not sustainable.

What is the maximum carrying capacity on our planet? We need to be more efficient in the use of our resources. It is important not to destroy natural landscapes. The aim would be to maintain the footprint we currently have and not enlarge it. Protecting untouched landscapes is important, but in many places like Brazil that does not work politically.

Education is a key factor. Places with educated women have lower birth rates. A key issue is migration from the rural to urban systems. What works in the city is complex and expensive. We live in a complicated world.

Q. When did you consider yourself to be a successful photographer?

I was 29 years old and the German magazine GEO hired me to do my first photo essay. Soon after that, National Geographic gave me my big assignment, and soon I had the freedom to work on my own ideas.

Q. Did you have any sponsors?

Sure! My sponsors were Christiana Breustedt, Geo's Director of Photography, and National Geographic's Director of Photography, Rich Clarkson, who launched the careers of many successful photographers.

I am in debt to those people as they gave me a chance. They were enablers for me.

In the photography business, people are proud to state that they "created" or "discovered" so and so talent, that they were the midwife for the careers of significant photographers.

You try to cultivate relations, going back to see people repeatedly. It is not as simple as it seems.

It is not like you get your big break and everything opens. It is an evolutionary process. You refine your skills and it is a feedback loop with a whole lot of little steps along the way. The enablers help you make your way up the food chain.

Q. There are certain milestones in our careers. What were yours?

I would say that an important milestone was when at 29 years of age, I landed my very first job with National Geographic. I went to see Rich Clarkson at National Geographic, who started a lot of careers in photography.

His predecessor was an also enabler, but difficult. The sign on his door was made by his secretaries. It read, "Please wipe your knees before entering." He was an old school guy from the Army Corps. When he fired you, he got out a little cap gun, and fired it into the telephone. Once you passed the quality test you were golden, but getting through that barrier was extremely difficult.

Q. Did you wipe your knees to enter?

(He laughs) No, I did not wipe my knees to enter. But I learned why I was not ready.

Q. What has changed in photography?

Many things, such as cameras have now advanced, are lighter, digital, electronic, so what was difficult or even impossible technically then is now easier. We all kept trying to push the envelope on what was possible.

The mobile phones have changed the things you can do. The people who came before me lived in the era of Speed Graphics. One shot; that is it.

Q. What questions should mentees ask a mentor?

In my profession, the mentors should ask the mentees how badly they really want to become professional photographers. You must want it bad. This is a difficult profession. You get many rejections, before you can make a living out of it.

If you do not love this with incredible passion, you are going on the wrong road.

Photography and journalism have been changing, the air is going out of the balloon. There is fraction of the jobs there used to be in both areas. Even though press freedom is of vital importance to democracy, jobs in journalism are disappearing. It is like in the oil industry, where it is harder and harder to find the big field, or produce more.

Q. Do you consider reaching 1 million followers in your social media an important milestone?

That was the result of a lot of work. I realized that Instagram and social media were overtaking print. For example, National Geographic has 2 million printed copies in circulation globally, but online they have 140 million followers. The worry with social media is about the business model for individual professionals, like me. We are worried about losing copyright to pictures. Then, I realized that Instagram was really becoming the dominant medium for images. More eyes see what is online than what is in print.

I started getting serious about Instagram and posting every day. I have to spend an hour or two on Instagram everyday. It is critical to having a voice.

It is not only the image, but the messages, the accompanying texts. That makes a difference. At the beginning, it seemed like a distraction, but now it is critical.

Instagram is somehow problematic because I have not found direct ways to gain revenue. As a business model, it is a challenging, but certainly a powerful tool.

Q. *Do you mentor others with your Social Media?*

When I was a young photographer, I used to look at National Geographic as the highest standard of what I wanted to do. I analyzed each page. I would go through so many pictures. What kind of lenses? I was deconstructing what I saw there. That was my guide. Social media has changed this approach. Now, people keep posting and follow their interests.

Now, I get people contacting me and showing me that they have enlarged my photos (from my Instagram!) as décor for their offices, and tell me, "You have become my inspiration."

With print media you did not get that kind of direct feedback.

Additionally, in social media, you can gauge your impact.

I cannot stress enough the importance of the picture captions. They are equally as important as the pictures. They are a kind of decoding. You can have a significant and loud voice.

Q. *As a Mentor what would be your key advice?*

To follow your passion. I followed my passion. I wanted to see the world from above.

It applies to all you want to do. For example, one of my children, a son, is passionate about French cooking, so, I encourage him. If you work hard, are smart and apply yourself, anything is possible.

Q. *When we initially contacted you, you told us you were not mentoring anyone. Why is that?*

It is the nature of my job. I do not need anyone to accompany me. It is not that I am anti-social. For documentary work you want to minimize your impact on the subject, so anything that is not necessary becomes a distraction. I must have all the gear in the car, and overseas I often have to have a driver and interpreter. If another person comes along, then I need another car, another hotel room, etc. It increases expenses, which diminishes field time.

What I fly is a one-seater paraglider. You keep it simple. I get requests from people who want to come and I'm happy to talk to people and give them advice, but I cannot take them along. It's just not possible.

I am happy to mentor others, but this kind of business does not easily allow direct mentoring in the field.

Q. *Is gender or age important in mentoring or sponsoring?*

I have not had female mentors as there were not many women advancing in this profession, when I was starting out. It was mostly white guys, although my first mentor, the guy whose trash I took out, was a Japanese American.

For me, race or gender is not particularly important per se. It is more about the picture. But I do see that different genders and cultures tend to see the world differently, and they bring that to their imagery, which is wonderful and important. What I always liked about photography, is you look at the picture first, not who took it, and either it works or it doesn't. It's a very level playing field, and people compete based on how they see, not on their gender, race, or income level. Now with smart phone photography and social media platforms, the barrier to entry is lower than it's ever been.

Q. *Do you consider that single mentoring sessions are also effective? In other words,* h*ave you had experiences with a one-chance encounter that made a difference?*

In field work, you go and solve problems in real time. That is the best. I am happy to advise people who want to show me their portfolio. I get impressed when someone develops a relationship. It's usually a one-encounter thing, but there are exceptions.

I remember a time when I was caught in a traffic jam in San Francisco with Ed Kashi. When I looked out the window, I saw Elliot Erwitt, an amazing photographer, in the car next to us. And so I got out of the car and walked over. I told him I was a big fan of his work gave him my card. He called the next day – and gave me a job as an assistant! I got to see one of my photographic heroes in action. The best training was always working as an assistant.

Q. *Do you think mentoring has to be in person, or are on-line, digital mentoring solutions equally effective?*

It is better in person. I enjoy the long drives and conversations of going out in the field on photography journeys. You learn not only how to do the photos, but also how to deal with people and what works to promote yourself. Being successful is more than taking pictures. Just like managing egos, it is a package deal

Q. *Did you enjoy any particular mentoring/sponsoring experience? Why? Tell us more about those joyful occasions!*

It is a wonderful when you can share those moments of success or failure with somebody else. Photography is indeed a bit like hunting. You wait, you find the right spot, only then you achieve something.

Q. Do you cherish any special photo of yours?

My favorites photos are those that were most difficult to do. Now, my best known photo is the one of camels from the air in the Omani desert. The one with the shadows. It went hyper viral on Internet and has been one of the best-selling photos.

I gravitate to things that are very difficult to do, and get access to. For aerial photos, the most memorable were in my motorized paraglider.

A Shared Selfie

- **Your favorite role model**: Georg Gerster, a pioneering Swiss aerial photographer. He photographed the whole world from a plane and there was always a story behind every picture. He was decoding the world.
- **One word to define your experience with mentoring:** Important.
- **One word to define your experience with sponsoring:** Crucial.
- **Who you wanted to be your mentor/sponsor, but you never had the chance to ask?** I reached out to Gerster. Sadly, I never got to meet him in his good years. When I went to Switzerland, he was very old, and we were unable to get together. I have photographed many of the same places he flew over, 30 years later.
- **Who would be an ideal mentor for you in this moment of your life?** Gerster in his younger years.
- **What would you like to ask him?** I would ask him how he selected or found the places that he photographed from the air. He did all his work from airplanes, and at that time it was much more difficult to find the places.
- **What do you think he would ask you?** He would ask me about my flying machine and if was it dangerous.

Afterthought

There were several coincidences occurred with the interview of Steinmetz.

The most outstanding one was that he was a Geophysicist, like both Eve and Maria Angela. It was even a bigger coincidence that he studied at Stanford at the same time Eve was there doing her graduate studies.

They chatted about their experiences with mentors on the campus, and George mentioned he left Geophysics and it was an amical separation. When he returned to campus recently as an invited speaker, he felt like the prodigal son.

Crafting the interview draft, Maria Angela marveled at two more coincidences: George Steinmetz shares her birthday, October the first, and on the day of the interview her automatic daily screensaver was a Steinmetz photograph.

Dr. Kamel Ben Naceur

"Receiving includes asking. Giving includes listening."

M. A. Capello and E. Sprunt, *Mentoring and Sponsoring*,
https://doi.org/10.1007/978-3-030-59433-6_5

A Glimpse

The energy sector faces many colossal challenges associated with changes to address climate change. Society currently depends on reliable sourcing of oil and gas to maintain and improve the quality of life of people around the globe. Modern society depends on electricity, water and transportation, all fueled by oil and gas. New developments in the sector are being shaped by sustainability goals.

Dr. Kamel Ben Naceur is the CEO of Nomadia Energy Consulting, one of the pioneering initiatives to address this huge sustainability challenge. Nomadia is "a new energy consulting company, focused on supporting capacity building and technical, project, and legal coordination for countries with greenfield developments." This demonstrates the leading role Dr. Ben Naceur is playing in the development of responsible and sustainable energy programs in emerging nations.

Dr. Ben Naceur's experience includes spearheading key programs with global, regional and national impact in top private and governmental organizations. He has served as the Chief Economist of the Abu Dhabi National Oil Company (ADNOC), as Director for Sustainability, Technology and Outlooks at the International Energy Agency (IEA), as Tunisia's Minister for Industry, Energy and Mines; and as Chief Economist at Schlumberger.

- He is a national of France and Tunisia
- He studied at the École Polytechnique and the École Normale Supérieure, Paris and worked at École des Mines as a research engineer
- 2022 President of the Society of Petroleum Engineers (SPE)
- In 1981, he joined Schlumberger and led the first business unit in the world to develop CCS (Carbon Capture and Storage). He became Chief Economist, and then President of Schlumberger's Technology operation in Rio de Janeiro, and Senior Advisor and Vice-President for technology of Schlumberger
- 2014 Tunisia Minister of Industry, Energy and Mines
- IEA International Energy Agency Director of Sustainable Energy Policy and Technology
- Chief Economist of ADNOC, the Abu Dhabi National Oil Company
- CEO of Nomadia Energy Consulting

A Personal Snapshot

Kamel is more than a colleague. He is a friend of both Maria Angela and Eve.

His support of colleagues, peers and especially younger professionals of all disciplines in oil and gas distinguishes Kamel. We were keen to interview him to learn how he developed his style of leadership and the differences between being "born a mentor" and "becoming a mentor".

The Interview

Q. Please tell us about your own mentoring experience. Let us discuss structured versus unstructured mentoring. How does structured mentoring adversely impact the effectiveness of mentoring?

In my opinion, structured mentoring that specifies a set of questions, timeline, and schedule, adversely impacts some of the components you want to see in mentoring.

There are individuals in your organizations who are rarely available and perhaps cannot follow such a structured approach. Also, personalities vary greatly from individual to individual, and not all—mentors and mentees—would thrive with or can follow a structured process.

Both structured and unstructured mentoring help. Structured is fantastic for the company in terms of overall workflows. Until 2000, when we completely reorganized the technical community of Schlumberger and created Eureka communities of knowledge with a technical ladder, one of the elements was mentoring of individuals. It helped a lot in terms of generalizing the mentoring concept.

Unstructured mentoring is very powerful and can address issues that may not fit into the formal structure. It can provide the mentees with the tools to meet or exceed their potential.

Once you start introducing different metrics and tools to measure progress, it removes some of the more natural components of the mentoring process that you want to occur. When you have the chance to work with exceptional individuals, then you don't necessarily want to work with a structured process.

Q. Do you have an example of either a structured or unstructured process for mentoring which was pivotal for success?

Yes, I will share one of structured mentoring. When I was in Research and Development, based in France, we had a very complex problem to solve. We were dealing with an issue pertinent to classic fluids and viscosity, related to some instability in the displacement.

So, we put together a team of researchers: an astrophysicist, a physical scientist, a mathematician and a geologist. They all had different backgrounds, but needed to communicate with each other and speak the same scientific language.

We assigned a mentor, a professor of polymeric liquids from Boston. It was incredible because before we shaped this team, they had been working for one and a half years, but working separately. In the first few months of working together under this professor, they achieved so much progress that they published their results in the very prestigious journals, NATURE and SCIENCE, that were the top publications in the world for acid fracking. Most of the credit was undoubtedly given to the fact that this team was mentored and guided throughout the process.

Q. *Were you there with them?*

Yes, I was there, and that is why I can testify that the on-going mentoring process was ideal and connected these brains and wills in a way that accelerated results.

Q. *Do you want to share an anecdote about mentoring that was unstructured and that was instrumental for you or your mentee?*

Let me think. (Dr. Ben Naceur paused for a while.)

When I was in South America, I had an experience very useful to exemplify this.

At that time, Schlumberger was structured into what are called "vertical" companies, and what we decided to do was to take a small group of individuals from each of the different companies, and put them together, to start thinking on a holistic way on behalf of our clients. We created the "Production Enhancement Group", and this worked so well, that it became the role model all across the organization. We did not impose any structure, but instead let the team organize themselves under our very loose guidelines.

My second example is more recent. It is a case from when I was a member of Tunisia's Cabinet as Minister of Industry, Energy and Mines. We were charged with developing a master investment plan, where we would invite key government representatives and investors, to support the needed rebuilding of the country. We were struggling to find the resources. One way to do it, was a traditionally travelled path that would involved asking renowned consulting firms like McKinsey, BCG and others to craft a plan with us, but we decided to go another way.

We went to all of the different Ministries, selecting thinkers, renowned talented strategists and planners, and we let them brainstorm with limited guidance to develop an in-country package that we would present to the investors. This was an extremely satisfactory experience, because when we presented the results of this work to the investors in September 2014, the investors, who were very impressed with the quality of the outcome, asked, "What consulting company did you use?" And we had not used any! It was the result of an unstructured or not very structured process.

Q. *Why didn't you want to use consulting companies?*

There were several reasons: One was time. We had a very limited time for this and didn't have time for the tendering of services. The other element was that we wanted to demonstrate we had in-country resources to do the work at stake with proper guidance.

It was a matter of self-recognition and creating self-assurance in a country that was about to be re-built. We needed that element. The guidance or mentoring was key.

Q. *Now, tell us about sponsoring. Have you been sponsored to advance in your career? And have you sponsored others?*

Sponsoring for me is an integral, an essential, part of mentoring. Mentoring is not only providing career guidance, providing support and helping on decisions, mentoring for me is also about playing an active role in different facets of the mentee's life.

One is in case of decisions pertinent to the mentee's career. Second, is by providing good words about that person inside the organization to promote him or her inside the corporate or organizational ladder, pushing the person upwards in the organization's ranks. I would say sponsoring is a subset of mentoring.

Q. *But sometimes sponsoring is a completely separate element. What do you think?*

For me, mentoring and sponsoring go together in the organizations.

I have worked in several organizations, and this has provided me a variety of cases of people whom I have recommended. For example, within SPE[1] and other professional societies, people who were endorsed by me have moved upwards very high and wouldn't have made the move if they were not recommended.

---

[1] SPE-The Society of Petroleum Engineers. Not-for-profit professional Society with more than 166,000 members in 156 countries.

But within the company it is a different story. We have the "high flyers", the individuals with an obvious potential to reach top roles, but we also have people in technical careers who do things well, probably within what is a much slower environment. Supporting the technical people is extremely important in order to make sure they achieve their best and ensure they are motivated towards their own goals.

Q. *Have you noticed cultural barriers in mentoring and sponsoring? Tell us your impressions on this.*

I have been very fortunate to work most of my career in a company where multicultural barriers were small. From the very top, there was a push to have cultural and national diversity. It was part of the company's DNA. Because that was an essential part of the fabric of the organization, all employees were used to working with people from all over the world. I did not experience cultural challenges in my different roles. It was a good thing.

This led me to advance changes in other organizations. When I assumed a role at the government of Tunisia, which was a very different environment as it was a mono cultural organization, I spearheaded dedicated efforts to encourage multicultural diversity in the administration. Multiculturalism is beneficial for decision making processes, and to increase the sustainability of an organization, as it enriches the perspectives, and helps in the decision-making processes.

Q. *During your tenure in the service company, was there a point at which diversity became more important? I am asking, because years, decades ago, there were very few women in operator companies, and probably fewer in oilfield service companies.*

For this case, I must share that again, we were very fortunate to have in Schlumberger at the CEO and human resources director levels forward-looking people who were pushing diversity and multiculturalism, when in other parts of the industry, it was not like that. Andrew Gould on one side, and on the other side, Pierre Bismuth. They had a fantastic open understanding of themes related to diversity.

And in terms of diversity, I remember that as managers we were inclined to have many mentees who were women. Many more than the share of the female component of our workforce. This was by design, because what we wanted to do was to fast track the women in the company. With this, they were very rapidly promoted up the company ranks. We had to do it by non-conventional moves and by major step-changes. We would move key people three, five or six positions up and needed to make sure that person would have the skill set needed. In that process, mentoring the person was extremely important to ensure her readiness.

Q. *What do you think about age and gender in relation to mentoring? Is it crucial? Important?*

I was very rapidly put in a management role. Most of the people I had to manage were of all ages. But I really didn't have the experience of being mentored by someone younger than me. I do not think is particularly important, and I would be glad if a younger person than me will mentor me in an issue or problem.

Q. *Kamel, were you sponsored in your career?*

Yes, in an unstructured way.

I had two main mentors in my life. One was Joe Mack, who is very well known in the industry. He was one of the inventors of the nodal analysis (Kermit Brown co-inventor). He was an academic, at the University of Tulsa. Joe was a mentor for me, but he was also a very good sponsor, who facilitated for me opportunities to move. He was tough. He was very well known to be tough. He was certainly one of the models I had in my career.

Andrew Gould was my other sponsor. Andrew became the CEO of Schlumberger. And he had this wonderful art of providing me with what I would call non-conventional assignments. He knew that those were initiatives I would love to do, rather than being in a much more structured role. He would assign me to those extraordinary opportunities, off the traditional loops!

For example, in 2002, Andrew asked me "*You know… I have heard about this CO_2 storage technology. Can you look into what can we do about it?*" I had carte blanche to go way beyond the company's normal line of business and start opening partnerships outside. And it was a fantastic opportunity for me, as it led me into an area that I cherished, the sustainability area. It also enabled me to gain insight about the overall energy sector, and way beyond that, into all different peripheral areas pertinent to the energy sector. An awesome opportunity indeed.

Q. *Would you like to tell us about your mentoring experiences at a global scale, perhaps pertinent to a matter close to your heart, the energy outlook or in sustainability?*

I think it is extremely important for everyone in this industry, to understand that we are a portion of a much bigger framework. I think every company should have an element of proactiveness, so that people, when they are at decision point in their careers have a good understanding of what lies ahead.

For example, in 2007 and 2008, I was advancing a program for top marketing in Schlumberger and was assigned a group of people. It would have taken me almost a year to provide this team with the tools that would help them to understand the power of marketing on the bigger picture. Part of that training was related to energy and energy climate. I included different components, like water and their interactions. That was one of the best experiences I had as a direct mentor.

Q. *Do you think mentoring has to be in person, or are the on-line, digital mentoring solutions equally effective?*

Effective mentoring must include interpersonal relationships. It must go way beyond the online tools, because every day is different, every person is different. In the past, the possibilities lying ahead were not as diverse. It really takes people with experience and bandwidth to decipher the future options and the strategic options for their case. The online tools are great but must be complemented by interpersonal relations. I cherish the personal contact, as I consider it is a powerful enabler of connections.

Q. *Some people had a special influence in our careers. Mentors and most particularly sponsors are some of those influential people. Please tell us about your own experience as mentee. You may want to go back into your childhood or high school years.*

I love this question! And I do have a story to tell you both.

I grew up with eight siblings. I was number seven. I had three older brothers. The eldest was a brilliant physicist, Raouf. The other, Seif, graduated from the France Polytechnic Paris, the top engineering school in France. In short, I had two inspiring role models. ahead of me.

When I went through the very difficult exams in France to get into engineering school entry was limited to a very few. I was fortunate that I was the first student from my country, Tunisia, to be admitted to the top two courses in France, the École Polytechnique and the École Normal Superieure.

Q. *Do your brothers look up to you now? Or do you still look up to them?*

I look up to them in many ways, and I am very proud when they ask me for advice. We enjoy a very fluent communication and closeness, which I cherish.

Q. *Is gender or age important in mentoring? And in sponsoring? Females mentoring females, or more-experienced professionals mentoring younger ones?*

No, I think though that more experienced people are natural mentors for younger ones. It is extremely important for me is to maintain a close connection with younger people. So, what I keep doing is teaching at the university.

Q. Oh! We did not know that. In which one?

At the École Sciences Politiques in Paris.

Q. What subject do you teach?

I am a professor for the master in International Energy program. I teach subjects pertinent to the nexus between energy and climate and water.

Q. For how long have you taught?

For three years now. My groups are composed of about 60-80 students, of very different backgrounds.

Q. What do you think about remote mentoring?

We are lucky. Our generation is blessed to count on LinkedIn, WhatsApp and other links that enable us to communicate widely, at a global scale, and in ways we did not imagine before, like video calls, or exchanging documents over the phone, recording messages instead of writing them, sending photos, and being vocal in social media. Like broadcasting to the world. Yes, maintaining a mentoring relationship is now much easier. You may even strengthen the relationship.

The mentoring relationship is not only about giving, but also receiving. Without feedback from both sides, the mentoring is useless.

Q. Getting feedback is critical for you or for your mentee?

I think is critical for both. It is about making sure that as a mentor you fully understand your mentee and learn from him. Many times, it may change your own perspectives on specific subjects.

You achieve the goals you set for the relationship by having the mentee achieve some of the steps of the different goals for the opportunities that lie ahead. But there is also the need to understand what didn't go right. Fundamentally, it is also a way of communicating back and acknowledging the successes or failure of the relationship.

Q. Tell us more about humbleness in mentoring.

Well, I would also put humbleness into the prequisites of the mentee and the mentor that they need to have as part of their personality, the notion of giving and receiving with humbleness.

Receiving includes asking. Part of giving includes listening to the other person.

Being humble from the mentor side is the attitude of being open minded and listening to the concerns and aspirations of the mentee.

Q. *What do you wish you would have done differently with regard to to mentoring or sponsoring, as a mentor or as a mentee?*

When I started, the tools that we had for communication were extremely rudimentary. The thing I regret is that we didn't have tools like we have today at the time I perhaps needed them the most. One of the most difficult things that happen to you moving from one country to the next in a career-long life at an international company, is maintaining relationships. Is very different if you work 30 to 40 years in a single geographical location. In this case, you maintain relations. But when you move around the world it is not easy; I was blessed to have a wife (Najoua) and three children who adapted very well to the changing environments.

Thankfully, things have improved with Internet.

Have I had regrets? I would say none that I recall. Perhaps the only thing is not having gone much more rapidly into a soft skills area, because the first part of my career was dedicated solely to improving my own technical skills. If I had complemented my technical skills with soft ones, I would have accelerated my career. I have provided this recommendation to mentees.

Q. *A question following up on this. What role have professional societies played in that connectivity among professionals in your case?*

I started in the industry in December of 1980. Ten months later, I had written my first SPE[1] paper. It was amazing to be immersed in the diversity of an organization like SPE. Throughout my career, professional societies have played a very important role.

Q. *Have you encouraged your mentees to get involved in SPE?*

Yes, for me it is a critical part of all my mentoring experiences. One of the first recommendations I provide my mentees is to have a more important role in the SPE or in their professional societies. I have been very pleased that some of the people I have recommended have achieved important roles in SPE.

Q. *Did you enjoy any mentoring or sponsoring experience? Why? Tell us more about those joyful occasions!*

With so many cases of mentoring, I think I could cluster them into three main categories:

First: Mentoring through management positions. I was very pleased some of the people I have mentored have gone on to very high levels in companies like Schlumberger and others. It is a recognition and acknowledgement that I have done a few things right.
Second: On technical side, I had the tremendous opportunity to work in an organization with a genuine technical ladder. I had people who moved high on Schlumberger technical ladder , what we would call the advisors and fellows.
Third: In professional societies, I have sponsored people who became members of the Board of Directors.

A Shared Selfie

- **Your favorite role model:** My top role model is my father. He was a lawyer, and I admired his ability to communicate and convince. I learned a lot from him about communication and negotiating.
- **One word to define your experience with mentoring:** I will answer with three: guidance, support and empathy.
- **One word to define your experience with sponsoring:** Influencing.
- **Who would be an ideal mentor for you in this moment of your life?** As this is an open conversation, let me share three ideal mentors I would like to have.

For leadership, I would look at the first Tunisian President, Habib Bourguiba. He had the courage to push ideas for the empowerment of women in the society decades ahead of his time.

For technology, my role model is the Egyptian, Ahmed Zewail, the first person from the Arab world to receive a Nobel prize in Chemistry. He worked at Princeton University. He received his award in 1999, and I met him a couple of times. He had an amazing perspective and you would be amazed at how extremely ethical this person was.

For role models, my third person is a lady, a winner of the Field Medal in Mathematics. She is Maryam Mirzakhani, the first woman from the Middle East to get a Field medal, which is the equivalent to Nobel prize for math.

I definitively like to think they could have been my mentors.

Afterthought

We finalized the interview in a very uplifting and positive mood. Kamel Ben Naceur has a way to make you feel at ease, and you do feel like if you have known him all your life.

We asked at the end why all his models were from the Arab world, since he had just praised the benefits of multiculturalism.

He told us that perhaps his list was focused, and limited, but that is common all around the world. We tend to admire those similar to ourselves, who have reached higher levels and been recognized in very special ways.

Kamel mentioned he would have included another role model, many more role models, and one that came to his mind was a French mathematician, Yves Meyer, who won the 2017 Abel award for his work on wavelet theory, that has been extremely important in geosciences.

Dr. Christine Ehlig-Economides

"Learning to listen has been fundamental for my mentoring."

M. A. Capello and E. Sprunt, *Mentoring and Sponsoring*,
https://doi.org/10.1007/978-3-030-59433-6_6

A Glimpse

Some people resemble lighthouses with a light that illuminates others and an internal steep and difficult spiral staircase which represents the hidden inner story of his or her ascent to leadership and recognition. Dr. Christine Ehlig-Economides' renowned technical work and broad life experiences motivated us to interview her to gain from her insights into her own challenges and numerous successes what she thinks is important about mentoring and sponsorship. We were keen in getting to learn from her experiences mentoring and being mentored.

We learned that like many other experienced professionals in the oil sector, she had not recognized the role these two key enablers played in her own career progression. Recently, she realized the importance of mentoring young professionals and was able to recognize key people who sponsored her and who were her mentors in life and work. For a person whose career spans several decades, these are wonderful perspectives and we were proud to be instrumental in capturing her thoughts on these topics.

As the co-author of a renowned book and author of numerous technical publications on a range of topics from well testing to hydraulic fracturing to completion designs, the expertise of Dr. Ehlig-Economides is widely recognized. She has a broad experience, that encompasses academia as well as service companies in oil and gas and has worked in many countries and cultures. She is one of the most admired women leaders in energy.

- Ph.D. in Petroleum Engineering, Stanford University, MS Chemical Engineering and Math Education MAT, University of Kansas; BA in Math-Science, Rice University.
- Professor of the Hugh Roy and Lillie Cranz Cullen Distinguished University Chair at the University of Houston, Petroleum Engineering Program, since September 2014, after several leading academic roles. Prior to that, was Professor of the Albert B. Stevens Endowed Chair of Texas A&M University.
- Global Account Manager (Schlumberger), Houston, TX, 1999–2003, Manager, GeoQuest Reservoir Technologies, Caracas 1997–1999.
- Other roles include Technical and Marketing Manager, Technical Advisor (Anadrill Schlumberger), Projects Leader (Schlumberger, France). Head, Petroleum Engineering Department (University of Alaska, Fairbanks, AK, 1981–1983).

- Co-Author of the textbook "Petroleum Production Systems", Second Edition, published by Prentice Hall, 2012.
- Recipient of many awards from the Society of Petroleum Engineers (SPE), including the recognition of SPE Honorary Membership in 2018.

A Personal Snapshot

Dr. Ehlig-Economides's career started along with that of her equally famous late husband, Michael Economides. Their highly successful dual careers inspired many others in the same situation (Maria and Herminio, her husband, are also a energy industry dual career couple.). Christine's light shines with such strength and warmth, that it is clear she was always a lighthouse, a bright one, that keeps illuminating other's careers and globally keeps generating knowledge useful for augmenting the productivity of reservoirs. We were delighted to include Christine in our compilation, as a friend and as an iconic woman in energy.

The Interview

Christine suggested some additional questions for her interview that were very illuminating. Her passion for sponsoring and mentoring generated a very rich and rewarding discussion. We hope you share our enthusiasm for her perspectives.

Q. *Please tell us about your own mentoring experiences.*

As a young woman, I was asked by recruiters in the USA many questions that have now been banned as illegal. I would be asked, if I was or was not married and how many children I was planning to have and when, and if my husband's career would take precedence. After many frustrating searches and interviews, I felt lucky to get a job with the Kansas Geological Survey studying multiphase flow, which I felt was an interesting job with plenty of challenges. During my student and early career employment experiences, I was not involved in mentoring. Actually, reflecting carefully about it, I have never been formally mentored at any time during my career. That was surprising for Maria, but not for Eve, who is Christine's contemporary. Eve was also not formally mentored until late in her career and was asked all of those illegal questions.

Q. *Can you tell us about your own experiences as a mentor? Have you mentored many people?*

Comparing industry and academia, today, I would say most mentoring experiences involve students at the university. I think that for me, mentoring starts as a casual conversation, perhaps, having lunch with someone. Maybe those were not labeled as formal mentoring sessions. Sometimes, I am labeled as a formal Mentor, and sometimes the conversations that take place are not even labeled formally as mentoring sessions.

I realized for me, mentoring occurs in a conversation and it is prompted by a remark of the other person or by me, or an observation, or by the mentee

More often, I think of mentoring as guiding young professionals in academia, rather than in industry or elsewhere. Maybe it is more common in industry now, but it wasn't at the beginning of my career in the corporate world.

Academics perceive mentoring as a mechanism for them to get over the hurdles in academia. This is especially true for those at the Assistant Professor level, because in 5–7 years, they must be recognized with tenure, or leave academia. Therefore, the first years of their academic career are very challenging and a critical time for them.

Careers in industry are not as dependent on what happens in the first five to seven years. The concept of "tenure" does not exist in the oil and gas industry. In return, people do not have the same concept of loyalty as they once did.

Q. *Christine, have you sponsored others in your career, and have you been sponsored by anyone?*

I recognize that I have been doing more sponsoring than I realized. I know several people who excel at this, and I would like to recognize the tremendous advocate that Joyce Holtzclaw has been in sponsoring women for SPE awards. Joyce has been instrumental in raising the awareness of the need to nominate women for awards. If you are not nominated, it is impossible to be selected.

I now see women finally reaching mid-managerial roles in their careers that when I was that age was unattainable. You need to advocate openly for deserving individuals so that they can receive promotions and awards. Now, I am in a position to often do something like that. This is sponsoring and goes beyond the scope of what mentoring can achieve in advancing careers and expediting promotions.

Q. *What do you consider is a primary ability to be able to mentor someone?*

Several characteristics are important in good mentors. I know people who are not mentors and simply don't want to be. I am sometimes guilty of that myself, as mentoring requires time and dedication. I consider most importantly, that you must leave your "ego" at home be flexible, and open to the experience. The mentoring experience should be an agreeable pact to both the mentor and the mentee.

Q. Are there any gender issues in the mentoring workflows? Should mentoring be female-female, male-male? Is it culturally acceptable everywhere?

Not so important. Ability to listen is paramount.

Q. What about role models? Is that a type of silent mentoring?

Being a role model is a way of mentoring, but quite different. Mentors may not be in a role you are seeking. Role models are valuable. When I was with Schlumberger, people told me that what I was doing was helping others. Looking back over the years, I now realize they were right, but at that time, I was skeptical. I do miss having had role models.

Q. Do you think mentoring must be in person, or are on-line, digital mentoring solutions equally effective?

In person works best. Phone or skype conversations are fine if the audio is good, and language comprehension difficulties are not excessive. On-line is harder and more subject to more error.

Q. Tell us about your sponsors. Has anyone sponsored you in your career?

My main sponsor was Steve Holditch. He sponsored me in several ways and numerous times. Steve as a leader and manager had many opportunities to sponsor people he trusted, and he used those. In the Texas Academy conference, he proposed putting me in a leadership position. And as people used to say about him (Steve Holditch passed away in 2019), "When Steve suggests something, it is very hard saying no." He would trust I could do a good job, and I benefitted from this advantage. People would approve his recommendations, selecting and endorsing my nominations for different leadership roles.

As for myself, I have not sponsored people as much, because I have not pursued the kinds of leadership positions that would enable me to sponsor others into leadership roles.

I do mentor people and suggest they take opportunities that would put them in a position in which they will be able to sponsor others. I guess this will qualify as sponsoring, too.

***Q.** Was your renowned husband, Dr. Michael Economides, your own personal mentor?*

Yes, it is good that you ask me this question, because Michael indeed, mentored and sponsored me all the time. In fact, I would say that Michael helped me more than Steve Holditch, whom I consider my best sponsor and mentor. But because Michael was my husband, I do not generally include him in this equation. Now that you ask, yes, Michael mentored many people all of the time and was very ambitious and leveraged that. He also mentored me professionally in many key and valuable ways. He wanted to be able to do mentoring for his most cherished students and liaisons and thrived in doing that. He was the "*Capo dei capi*" (the boss of the bosses, in Italian), and I of course enjoyed tremendously our partnership in work and in life.

***(Comment)** It was refreshing to see Christine opening up about this decades-long partnership. Then, Christine talked extensively about the Robert Frost's poem referring taking the road less traveled. She said that several lines from Frost's poem were the theme of her life. She consistently picked paths that are not the most traveled and that led her in unexpected directions. She said, "When role models and paths are not there, you must take chances."*

She mentioned as an example how she decided to work in Alaska. It was an unexpected option, that came according to her "out of the blue," but that made sense to Michael, who proposed it to her, and she accepted without hesitation. People would ask them the whys and the basis for that decision, but they took it on the spur of the moment to experience something wild and new for them. They never regretted it. Christine said her life has been like that and she is not sad about it.

***Q.** How do you initiate your mentoring sessions? Is online mentoring good?*

You want a quiet place. Not a crowded restaurant. My mentoring discussions have been in quiet settings, always.

About on-line mentoring, problems can occur, generally because of comprehension difficulties. What I like in academia and in industry is diversity, but when you are working with someone from a different background, you may have language comprehension difficulties. Visual contact is very beneficial. When there is only a voice-line, you cannot see the body language. Even in person, language can be an issue. This happens a lot for me.

Online mentoring conducted solely through written emails have a serious disadvantage. We all know written words can be misunderstood. But there is a record, and oral conversations can be misremembered. The written record is the primary advantage of online mentoring.

Q. What kinds of questions do you think people need to ask in a mentoring session?

Especially for professors looking forward to gaining tenure, there are certain critical questions, but many times there are no prescribed questions. Do I have suggested questions? No, and I have never encountered a problem with people not knowing what to ask. In general, I have had to deal on the fly with easy or difficult questions from people who seek my mentoring.

Q. Do you think a mentoring session must be structured?

I like to let the conversation flow, but it would be good to have workshops to teach how to mentor and how to receive mentoring. This would greatly improve that exchange.

I remember what a banking woman said about a promotion that she thought should have been offered to her. When she got the nerve to go and ask her boss (a woman), "Why didn't I get this position?" the boss answered, "I didn't even imagine you wanted it." The message is clear, and it is "MAKE IT KNOWN." Get people to know what you want, especially those who are in a role with the capacity to influence decisions related to your promotion or to directly promote you.

There are situations where mentoring could bring make that happen. I am learning more about the formality of the academic environment due to the involvement I have at University of Houston. But for sure, in academic and in corporate environments, we must follow the rules of the game and understand who the influencers are and who are our possible sponsors. Mentoring can help in identifying identify potential sponsors.

Q. Have you been mentored by a younger person?

Yes, if you want to get things done, you need to relate to younger people who are talented. It depends on the topic. Again, leave your ego at home!

Q. Do you have any cherished moment related to mentoring?

Probably the most appreciated mentoring I have done has been in the last five years at University of Houston. Women's promotions in academia are a fascinating case to ponder, with regard to the importance of mentoring and sponsoring and the need for role modeling.

A Shared Selfie

- **Your favorite role models**: Michael Economides and Steve Holditch.
- **One word to define your experience with mentoring**: Willingness.
- **One word to define your experience with sponsoring**: Accomplishment.
- **Who would be an ideal mentor for you in this moment of your life?** There is a woman I met not long ago, in my choir activity, that has provided me valuable insights on life and work. Her name is Rochella Cooper. She is for me now the ideal mentor. I cherish her points of view.
- **What would you ask this person?** What could be worthwhile goals?
- **What question would you like this mentor to ask you?** Are you sure you want to do that?

Afterthought

After the interview, Christine mentioned to us that she wanted to highlight that she often sees people confuse leadership with position. The latter brings a title and can enable sponsorship. The former is possible in many ways. SPE has recognized Christine many times through the years even though she has not held high positions. Technically, she "has not changed the world." She thinks her future Distinguished Lecture should be on the "changed the world" category. Even so, she considers her future lecture would need to highlight the leadership that already exists. Rarely is there anything truly new in this world, and luck is about recognizing and seizing opportunities that confront you.

Dr. Luis Augusto Pacheco Rodriguez

"Only results lead the way towards continued success."

M. A. Capello and E. Sprunt, *Mentoring and Sponsoring*,
https://doi.org/10.1007/978-3-030-59433-6_7

A Glimpse

Generation X and Baby Boomer managers and executives are not keen on participating actively in social media. But those with an avid desire to communicate have realized the tremendous power provided by Twitter and Instagram to share opinions, argue one's stand on an issue, or communicate a point of view and use them widely. Examples from all sorts of political perspectives are very visible and known. Dr. Luis Pacheco has a solid imprint in Twitter and is leaving his mark and inspiring followers around the world

We were pleased to have the opportunity to interview one of the leaders of the PDVSA who led an era of excellence. That era shaped thousands of professionals who migrated in search of work elsewhere after the dismissal of 22,000 employees in 2003. Today, Venezuelans who left PDVSA, not because of the massive layoffs, but due to the deterioration of the company and their country, comprise a recognized diaspora of excellence in oil and gas, and you can find Venezuelans working in oil and gas in every country on the planet.

The Venezuela oil industry is at a very low point, unlike any in its history. What happened with its processes, performance and values will be included in case studies of business schools for decades. The country with the largest reserves of oil in the world, surpassing those of Saudi Arabia, witnessed a decline of its oil production from about three million barrels of oil per day to less than one.

Dr. Pacheco is highly knowledgeable about the causes of this tragic destruction of what was once a model of corporate excellence, and he was selected as the Chairperson of the Ad-hoc Administrative Board of PDVSA [in exile].

Our interview focused not on the sad situation or his plans. It was about mentoring and sponsoring. Incredibly—as we did not expect it—he warned us he was not the best person to interview on these topics. He did not consider himself an experienced mentor. In retrospective, we are very glad we interviewed him, as this turned out to be one of the most enlightening interviews about sponsoring limitations and pitfalls for corporate progression.

- Chairman, Ad-Hoc Administrative Board of PDVSA since May 2019.
- Non-resident Fellow Baker Institute, Rice University, Houston.
- Previously held top strategic roles in Petroleos de Venezuela (PDVSA), as Executive Director of Corporate Planning, or Chief Strategist, concurrently with Head of the office of the President, at a time when PDVSA was a vertically integrated Energy Corporation with operations on three continents and revenues in excess of 40 billion USD and more than 40,000 employees.

- Visiting Lecturer at Universidad Sergio Arboleda and UNIANDES (Bogotá), IESA (Caracas), Kennedy School—Harvard University, Georgetown University—Washington (USA), Tecnológico de Monterrey (Mexico).
- Former Senior Vice-President of Strategy and AIT, Pacific Exploration and Production, Bogotá Colombia, producing 310,000 BOPD.
- CEO of BITOR (Venezuela), producing and exporting in excess of 100 MBOPD of extra heavy oil.
- Exploration and Production Planning Manager of PDVSA, with 23,000 employees and 5 Billion USD budget.
- Senior Advisor for the Board of CANTV, the most important telecommunications company in Venezuela on issues related to energy, strategy formulation and corporate planning.
- Founder of his own consulting company that developed projects for Shell, PEMEX, REPSOLYPF and Citibank, CAF (Corporacion Andina de Fomento), World Bank and IADB (Interamerican Development Bank).
- Ph.D. in Mechanical Engineering, University of London, Imperial College of Science & Technology (UK); M. Sc, in Mechanical Engineering, University of Manchester (UK); and Mechanical Engineer from Universidad del Zulia (Venezuela).

A Personal Snapshot

Dr. Luis Pacheco is an example of a leading Venezuelan professional who has overcome difficulties and propelled positive change. His career is a compendium of incredible coincidences that, combined, enabled him to climb the corporate ladder rapidly. The time for our interview was insufficient to address everything we would have liked to learn. We finalized the interview yearning to know more, to laugh more, and to learn more.

The future of Venezuela depends on the health of its oil industry, which is now sick and in need of urgent care. Recovery must address, among other things, the rehabilitation of PDVSA. Current production levels show that PDVSA has been crushed by legal issues and is ethically compromised to unthinkable levels. Luis Augusto Pacheco-Rodriguez is the man selected to be the chair of the Ad-Hoc Board of Directors of a proposed reformulation of PDVSA. Very conscious of the importance of the responsibility this task would entail, we were pleased he shared some of his time for an interview.

His time was very limited, so we started immediately with our interview questions.

The Interview

Q. Please tell us about your own mentoring experiences. Have you mentored many people? Do young professionals reach to you to receive mentoring? How does your mentoring process start?

I am not necessarily a good person to be interviewed about mentoring. Not as a mentee, not as a mentor. I have never been a sponsor or mentor. Well, no, the sponsoring denial is a lie. I have sponsored and been sponsored, but believe me, most people who tell you they have mentored many people are probably lying.

Not the sponsoring element, as I do think sponsoring is a crucial element in a successful career. I am mainly referring to mentoring. Most people, who tell you that they have been mentors, in my opinion, are lying or kidding themselves. Nowadays, mentoring is a popular activity, the politically correct answer.

So, I prefer to discuss the sponsoring element of your interview, which for me is more interesting, much more precious.

Sponsoring is not a bad thing. That is how human groups behave. People like certain kinds of people. Generally, their images, their copycat selves. People love sponsoring others who are obviously like them, who laugh at their jokes because they resonate; similar in values, in perspectives, in culture, in upbringing. That creates confidence. A connection.

They sponsor people whom they feel are just like them. In a certain way, they end up sponsoring themselves.

Maria, if you remember from your time in Venezuela, you were probably told, or even taught, that Maraven people behave like Maraven, and Lagoven people behaved like Lagoven. You were probably told that one could tell the corporate heritage of each employee by their working styles. It was argued that because those corporate cultures were so incredibly strong, they ended by transforming the employees until they fit the norm and were tailored to a specific pattern.

In Venezuela, you would be told that Maraven people were more informal, not good at projects, but very creative. That Lagoven people, instead, were very budget-minded, and not original or flexible enough. These are descriptions made by whoever was not a member of either one of these two corporate cultures, Maraven (ex-Shell) and Lagoven (ex-Exxon). These were also very practical and summarized stereotypes. Then, there were also the styles of talking and even dressing. There was even the "Lagoven" suit, for example.

My conclusion is not that those organizations were so strong as to modify personal traits, but that people in power positions would pick those who were similar to themselves. In recruitment, and also for promotions. Maraven people were not any more successful in grooming the personality of its employees. They both were extremely good at choosing people who were similar to themselves. They maintained or preserved the culture not by a forced selection of the best, but a natural selection of similar ones. And this is not a natural selection. If you think about it, it is a way of sponsoring. Perhaps, sponsoring is a nice word for "Gang-like" behavior.

It is all, I believe, related to how human groups behave. People are naturally drawn to people who resemble them. They end up being a sponsor because, in a way, they are sponsoring themselves.

I started working in the oil industry, not in Venezuela, but the Netherlands, with Shell. I worked at their Research Laboratories, where a mixture of British, Dutch, French, and Venezuelan employees teamed up to do the work. The culture of Shell was not as interpreted in Venezuela to be the one of Maraven. It was very rigid. They did not like undisciplined people (like me, just to reinforce the point). People claimed it was unnecessary to learn Dutch to work at Shell, but clearly, those who learned Dutch had a better career than those who did not, especially in exploration and production. Exploration and Production (E&P) were in the hands of the Dutch, and Finances and Planning were in the hands of the British—a generalization to be sure. Again, self-selection and a customized selection process could be observed.

When I returned to Venezuela, I was naturally drawn to Maraven, not because I was coming from Shell. I noticed they were more akin to my personality.

I received three excellent offers from the three different PDVSA affiliates, but I was drawn to the way the people in Maraven talked to me during the interview process. I was attracted since my childhood to Shell and then to Maraven. Maybe because my grandfather was in the engineer in charge of digging the trench that contained the famous Barrosos 2 gusher that brought the world's attention to Venezuela's oil in 1922.

All these things matter. Even if you do not realize it at the time.

I must also disclose that when I finalized my first degree, I went directly to graduate studies. I wanted to study and work in things different from the oil industry. For me, it looked that it was too much hard work. Yes, all that work in the field was not for me.

Q. *For you, Luis, which is more important, mentoring or sponsoring?*

In my experience, sponsoring is more important than mentoring. However, sponsoring may originate in personal empathy, but in the end, any endorsement necessarily must be supported by results.

People are drawn to people who get the job done for them, to people who make them look good. Managers may select or recruit the odd nephew or far away friend of the family, and perhaps the quick-to-smile engineer who seemed to attract everyone to his cheerful personality. In the end, however, any endorsement and sponsoring must be reinforced by results. The success of the sponsored is not equivalent to altruistic behavior. There might be empathy at recruitment phases, but only results lead the way towards continued success.

Q. *Did anyone sponsor you?*

I probably was, but not in an explicit way.

Lucky coincidences happen all the time. I believe luck matters, as well as the utilization of skills. There is always a risk. When I left Shell- on a whim, if you like. I did it because my boss and I did not have the same perspective about what my career should look like.

When I went back to Venezuela, I went to a specific project in a refinery. I immediately felt that I had made a big mistake. My job was in Cardon, a God-forsaken place where I was ill at ease. Fortunately, for me, the project was cancelled. The oil price went down, and it was no longer economically attractive. The manager there told me I should not worry, as he had the perfect job for me in E&P Planning at the central office.

So, I was sent to Caracas, and the boss of E&P Planning said he didn't have a job for me. "Who told you I had a job for you? Who sent you here?" I explained that Luis Giusti[1] had sent me. And he repeated, "Well, anyway, I don't have a job for you."

At that very moment, the phone rang, and it was a guy from one of the other planning departments, Economic Evaluations. They were urgently looking for a substitute for a fellow who was unexpectedly going on vacation. I was there, available, in front of him, hence, my new boss realized it was a perfect opportunity, and so I was sent there. It was definitively a strange circumstance, as I had been a researcher in offshore structures, and got a job in Planning for E&P, a post about which I did not know anything.

My new boss asked me "Have you ever done an economic evaluation of a project?"

[1] *Luis E. Giusti Lopez (1944–), a Venezuelan Oil Executive, who was President of PDVSA 1994–1999.*

A laconic "no" was my answer.

Without hesitation, he just sent me across the street to a bookshop, called "Tecni-Ciencia" telling me to buy a book on economic valuations and to study it. I went to purchase that book. And that started my career in Planning.

After the vacation, the fellow for whom I was substituting got transferred to another job. I stayed in the economics evaluation section of planning, never returning to the original post intended for me in Caracas, which was in Project Engineering. I knew how to program in what at the time was the famous and sophisticated Lotus123, and I ended up programming their worksheets, which led me gradually to become manager of economic evaluations.

And that is not all. More coincidences happened along the way.

They taught me how to do economic evaluations. I did not know anything about oil reservoirs or how much a well cost. I was taught or had to learn on my own. Learned the hard (maybe faster?) way, of having to go and ask people about doubts and facts. And people were willing to teach you.

I also learned about human nature and the hidden art of understanding social interactions. I eventually discovered that if one took along a pretty and intelligent woman to the meeting with the big boss, she was a "secret weapon". He would be less stubborn. It worked many times.

And I learned more and more about the complexities of planning in a large oil operator. Hard, technical, as well as soft skills. More things happened along the way. I created confidence in my peers and supervisors by doing my job consistently. More people started thinking "Aha! This fellow is reliable".

***Q.** Have you sponsored others?*

Yes, I have sponsored others and have also made some mistakes doing that as well. People who laughed at my jokes and whom I thought were reliable and intelligent. I like people who have interests beyond their immediate job in the oil industry. I have sponsored some people who were mistakes, who ended up being part of the Chavez-related corruption gang.

***Q.** Did you receive informal mentoring?*

I would say no. When I started, I was just told how to do economic evaluations. Because I had minimal experience and was in deep water, my peers and supervisors taught me, explained to me—about reservoirs and well costs. I had to learn a lot on my own. One must be willing to do that.

This approach worked well in Maraven, where I was employed. If I had worked in Lagoven, I would not have done that. The system there was a lot more structured. I once gave a wrong answer about deferred oil production, and still to this day, 30 years after my mistake, some ex-colleagues who were working in Maraven, remind me about that.

Such was the culture of the company.

Q. *When you were finding your own way in Planning, did anyone explain to you who were the important people?*

Well, there were many things to learn. But did they tell me precisely who the important people were? No! Absolutely not! If you are not a fool, no one should need to tell you that.

Q. *In your own experience, what are the questions that the mentees should ask you? Or in other words, what questions do you want to be asked in a mentoring session?*

Tough question. You learn from your mistakes. What are your major mistakes at the end of a long career? Did I talk out of time or contradict the boss? I tend to be very controversial and contradict my bosses. This conduct is definitively not something I would recommend to anybody. On the other hand, I like people to contradict me; it is a good sport.

I have been asked many times if getting an MBA is mandatory or recommended? In these cases, I counter-ask—are you going to be happy doing it? Do you like to think or are you doing it to raise your salary? Is your wife happy about it?

I have not had many women working for me. I have had women peers, and that is a different challenge.

Q. *Do you think we need more mentoring on soft skills?*

Do you need to be taught about that? When we leave school, we lack three critical skills: raising children, doing personal finances, and how to negotiate. You learn by mistakes. Living with children should teach you how to live with adults. Almost everything is a negotiation structure. Life is all about interaction; interacting with people.

Q. *When did you realize the importance of soft skills?*

I think I recognized that I was not very good at negotiating early on. You learn to do your work correctly by making mistakes; I cannot stress that enough. I have often been told that I am impatient with fools. And yes, I am; with people whom I think are fools, not that they are. Perhaps I am more aware of it now than when I was younger. You think people do not understand you because they are fools, rather than you do not know how to explain yourself. We need to learn how to connect with other people.

It is easier, much easier to mentor younger people. I was once a lecturer at university, and I loved doing that. I think this is the right moment to confess that I wanted to be an actor. Being a teacher is as close to being an actor as you can get. Moreover, a presentation is a negotiation with the audience.

Q. *Do you think gender or age is important for mentoring and sponsoring?*

I think gender is not important, but age is. The age of the mentor is important—you need to be at a particular stage in life—you need to be at a point where you are willing to be mentored or to serve as a mentor. When you are older, you are almost impermeable.

Q. *What do you wish you would have done differently in reference to mentoring or sponsoring, as a mentor or as a mentee?*

I think I should have sought sponsors earlier. But I don't know. I'm not sure.

Q. *Did you enjoy any particular mentoring or sponsoring experience?*

Sometimes, when in a meeting, presenting or discussing important issues with my team, I would like to tell them, "I will now go in mentoring mode", and just preach to them! I would extract from past experience what would relate to the particular work and used what I learned from a boss or from what I learned in the past.

Q. *How do you envision your future?*

I am currently the Chairperson of PDVSA in exile. It is a real challenge. If things go correctly, we will have a short time in which to do anything.

I ask myself whether the new generation even considers this concept that we are dealing with, because of how often they change jobs. People nowadays spend less time in a job than our generation.

I cannot but reflect on root issues, like how do we fix the Venezuelan industry? What to do when the mentors and sponsors are from two generations ago. Do their concepts and ideas still apply? Venezuela is an entirely different and complex case because you have not had continuity; you have skipped generations of knowledge.

Disruption was incommensurable, and recovery will be a rocky and uphill road.

Moral principles have been affected. How much have recent times changed societal values? How do you mentor or sponsor people who have grown up with a different mindset? With different values? How do you build corporate ethics?

I have a question for Maria: can some work or research be done about cross-cultural mentoring/sponsorship to help with the Venezuelan situation? I think mentors will have to come from other cultures.

A Shared Selfie

- **Your favorite role models**: It would have to be a composite for work and life. I would say Alberto Quiros Corradi,[2] a kind of Renaissance man who was politically savvy. One of the very few oil people in Venezuela with practical political understanding. Then, it would have to be Carlos E. Castillo, a brilliant guy with a fine sense of humor. I would also pick Luis Giusti, who to me is a composite of those two guys.
- **One word to define your experience with mentoring**: Imitation—trying to make a copy of yourself.
- **One word to define your experience with sponsoring**: Betting. Betting on the future or your future.
- **Who you wanted to be your mentor/sponsor, but you never had the chance to ask?** I was lucky enough to have been exposed to very good people. Again, that is pure luck.
- **Who would be an ideal mentor for you in this moment of your life?** Leonardo da Vinci, but probably he was not a good mentor. So, I would choose someone very unattainable, like Richard Feynman.
- **What would you ask this person?** How does one remain human through difficult circumstances?
- **What question would you like this mentor to ask you?** What mistakes have I made?

Afterthought

When we asked Luis Pacheco about his favorite role models, his answer opened into an extensive list of possible mentors and role models. We confirmed we had a curious soul in front of us, when he mentioned very different role models in his passion for reading.

[2]Alberto Quiros Corradi (1931–2015), was a Venezuelan Corporate leader and politician. Member of the Board of PDVSA, President of Shell Venezuela, Lagoven and Maraven. President of newspapers El Nacional an El Diario de Caracas, and President of Tamanaco Hotel.

He told us that he was at the time of the interview influenced by the writings of George Orwell, because of the very economical and succinct style with which he writes. Also, he was reading the biography of Leonardo da Vinci and was passionate about the life of Albert Einstein.

Luis was humble, as he expressed in melancholic tone that he wished he had the range that those people had. "I don't have that range."

We instead think he is already in a position able to rescue Venezuela from the people who have derailed Venezuela from an ethical and successful path. He will most probably be instrumental in redirecting the future into fruitful and healthy growth for Venezuela.

And we will cheer and remain hopeful for that to happen.

Riccardo Piatti

"You cannot buy experience ."

M. A. Capello and E. Sprunt, *Mentoring and Sponsoring*,
https://doi.org/10.1007/978-3-030-59433-6_8

A Glimpse

Spectacular. This is one of the most frequent adjectives applied to tennis, a sport that is played on all kinds of surfaces by millions of people and is second only to soccer in popularity. With a simple scoring system (15, 30, 40, game), tennis has produced exemplary athletic role models including Novak Djokovic, Rafael Nadal, Roger Federer and Serena Williams. Former tennis stars Martina Navratilova and Jimmy Connors are engraved in our collective memory as legends.

Tennis stars are ranked using a merit-based system established by the Association of Tennis Professionals (ATP). In sports, statistics matter a great deal. Trainers who can improve the rank of players receive attention. Riccardo Piatti is an Italian tennis coach who has excelled in improving the ranking of players. Players aiming for excellence seek his assistance. He has coached several players ranked in the top 10 by the ATP, including Novak Djokovic, Ivan Ljubičić, Richard Gasquet, and Milos Raonic.

Piatti trained the current number 1 ranked player in the world, Novak Djokovic, when he was 17 years of age. He is the coach behind the young Italian Yannik Sinner's sky-rocket ascent from ATP rank 670 to 78. Between 1997 and 2012, he propelled Ivan Ljubičić from an ATP rank of 954 to his career-high ranking of number 3. He pushed Milos Raonic from ATP rank of 11 to his career-high of number 3.

- Born in Como (Italy) in 1958.
- Piatti began playing tennis at the age of nine at the tennis club of Villa d'Este in Cernobbio, Italy.
- When he was twenty years old, he replaced the Villa D'Este head coach.
- At 25, he moved to USA, to learn to coach at the Nick Bollettieri Tennis Academy in Brandenton, Florida.
- Became an instructor in 1982, after attending the National School of Tennis Instructors, becoming manager and captain of the U16 team of the Italian Tennis Federation between 1984 and 1988.
- He became a private coach for professional players in 1988.
- Among the early players he coached are Renzo Furlan (career-high world No. 19), Cristiano Caratti (career-high world No. 26), and Omar Camporese (career-high world No. 18). They were collectively known as "Piatti's boys".
- Since 1990, trained Ivan Ljubičić, when he was ranked No. 954 in the world. While working with Piatti, Ljubičić achieved a career-high world ranking of No. 3, winning 10 ATP tournaments and a Bronze Medal at the Athens Olympics in 2004.

- From fall 2005 until June 2006, he coached Novak Djokovic who was then 17 to 18 years old. As of April 2020, Djokovic is the ATP top ranked player, ranked No. 1.
- He coached Richard Gasquet's rise from No. 31 to No. 9.
- In 2013, Piatti began working with Milos Raonic, who was then No. 11 and propelled him to a career-high of No. 3.
- Five times Grand Slam champion Maria Sharapova began working with Piatti in 2019, before she was injured in the Australia open.
- He opened his own academy in Italy, the "Piatti Tennis Center" in 2018, offering services to top players with personal coaches, and athletic and mental trainers of international renown.

A Personal Snapshot

Riccardo Piatti's insights on mentoring and sponsoring are based on his experiences coaching top tennis stars with different nationalities, personalities and levels of skill.

We wanted to know the secret to his success. We wanted to learn what if anything was similar between coaching star athletes and mentoring talent in the energy industry. We found many similarities.

Piatti acquired his mentoring skills in a very competitive field. He has greatly enriched his profession without losing his kind-heartedness and concern for his protégées. That aspect enchanted us.

The Interview

Q. Please tell us about your mentoring experiences. You have mentored tennis stars, but how did you get started?

I am a coach of tennis talent, especially those who are young, and I have a tennis center where I do this. The most recent is Jannik Sinner, an Italian tennis player who at 18 years of age won the Italian "Next Gen" open, a prestigious world championship for those under the age of 21.

When I was 21 years of age, I announced to my parents that I loved tennis. My father was an industrialist in Como, Italy and my mother a teacher. I started by researching the sport because I wanted to understand the intricacies of the game. In my search for knowledge about the game, I realized there were missing pieces and that I needed to get to know myself better. If I did not know myself, I could not deliver any technical messages.

From 1981 to 1982, I attended the School for Tennis Coaches in Italy, which at that time was under the direction of Antonio Rasicci. I learned a lot, but I felt that the more I advanced, the less I understood. So, following the recommendation of Gianni Clerici, a famous trainer and former tennis player, who was from the same region in Italy as me, I decided to leave Italy to pursue new horizons and went for several months to the prestigious training campus of Nick Bollettieri in the US. Upon my return to Italy, I was offered the role of Vice-Director of the School for Tennis Coaches.

When I reflect on my own story of mentoring, I realize I have been very lucky to have the opportunity of coaching so many different personalities, nationalities and cultures. This gave me experience and perspective which provides the foundation for my knowledge of the game.

I adapt to the people I encounter in my path. My ability to teach, to mentor, to guide, to coach, is grounded on knowledge of myself, I aim to get to know the person in front of me his or her education, upbringing, values, style, and personality.

My mentees are generally very young stars in tennis or professional tennis players. When I work with a young talent, for example 22 years old, it is pivotal for me to understand the kind of upbringing he received. If the parents were engineers, then most probably the education was very rigid, systematic, and I will not succeed if my approach is naïve. I need to use a direct communication style that is to the point. I must be clear from the beginning in a mathematical style.

Communication is essential for success. I must find the right way to explain the game and what they need to accomplish in their technique in ways they will grasp, internalize and embrace.

If the individual has had family issues of any kind, my role changes and is more maternal. I had a case of a young fellow whose mother was very insecure, and he lacked affection. I had to offer my communication with a caring interest and sincerity.

I adapt to the individual I am mentoring, so that he may understand.

Q. *How do you handle communication with your own team? I understand that every star you coach requires a full team of support. Tell me more about this.*

The second phase of coaching a star player or promising talent requires more than just a single coach. I organize a team of support people for each athlete. The support technical team is composed of a physical trainer, an osteopathic doctor and a physiotherapist. In some cases, a mental trainer/coach is needed as well. I am the coordinator and leader of this team and bridge across the specialties of each person in the support team with a sole aim, which is the optimal preparation of the talented person.

Sometimes, I am a translator, from one discipline to the next. My work is heavily based and dependent on them. I act as a leader of the team.

Q. *Do you want to share an anecdote of a case in which the support team played a key role?*

I had a tennis talent, a very talented individual and good person who was the son of two engineers. For a specific important game, I had instructed and agreed with him on a certain strategy. Then I noticed he would go to the tennis court and do something else! He would speak to me in his own language with a bad attitude. I did not understand what it was that he was trying to tell me.

He said, *"Excuse me coach, but when I am under pressure, I can't understand anything."*

I had received complaints from his support team that he treated them badly, that he had a bad attitude and was disrespectful. I interpreted these complaints as signals. I advised him that we needed to find a person to help him with this problem, which was affecting the effectiveness of his coaching and the quality of his playing in tournaments. He told me that he was not angry with the team or with me. It was that when he was under stress, he would see all red.

From that point onwards, I collaborated with Professor Beppe Vercelli, who is the mental coach of the Juventus football team, the top Series A team in Italy and one of the top teams in the world.

The mental preparation of an individual is one of the most important elements in my holistic approach to coaching tennis in highly competitive environments.

Q. *And your own mentoring? Did someone mentor you?*

When I was certified as tennis coach, my first job was as a coach for the Italian Tennis Federation. At that time there were no personal tennis coaches in Italy. In the entire world, personal coaches were rare. In the US, Björn Borg (a record setter player, number 1 in '77,' 78, '79 and '80) had a personal coach, but not many other players did.

My goal was not to become a personal coach. I was engaged in the profession when the main coach at the organization for which I was working was injured, and I was asked to substitute for him.

I had teachers and mentors who explained to me how to become a coach. For instance, in 1986, one of those mentors, Vittorio Roiati, was based in the city of Ascoli Piceno in Italy. He would rarely leave his city, going out to Rome only once a year for the championship played in the capital. He was a fanatic of the game and watched every game of important players on TV, analyzing them in detail.

We met at the Roma tournament in 1986, and he brought to my attention a very young player, who was 16 years old. He emphasized he would be a champion. "*Why*?" I asked. He replied that he had been watching his movements. When that teenager served the ball, he would move with three steps that were almost identical to those of Mikael Pernfors (one of the top 10 tennis players in history). His style in the back was also like Pernfors.

That 16 year old was Andre Agassi. That teenager was to become the world number 1 in 1995 and 1999, an eight-time Grand Slam champion and a 1996 Olympic gold medalist. The first of only two to achieve the Career Golden Slam (career Grand Slam and Olympic Gold Medal, the other being Rafael Nadal).

That year, the first after becoming a professional player, Mikael Pernfors went to the final match in Paris, the Grand Slam Roland Garros. It was the same day I went to Sarasota, Florida, to work with the coach Nick Bollettieri, who was pioneering the concept of a tennis boarding school.

When I was there, I met Andre Agassi.

I asked Bollettieri, "*How does Agassi learn how to move like that*?" He told me, "*because he analyzes Michael Penforst's videos every single day.*"

I realized that very moment, the huge caliber of Roiati, who had already envisioned that quality, and the style of Agassi at his home, in Ascoli Piceno. Destiny is a wonderful thing.

Vittorio Roiati did not speak English. He didn't know the network of influence in the USA, but he was looking ahead of everyone else. He helped me a lot in my own career, and I mean a lot. To be a successful coach, you must see what will happen in the future.

In sports, there are innumerable variables that shape a career in critical matches, like misfortunes, injuries, health, climate, nerves, and many more. A good coach must be able to envision the future in his mentees.

Q. *Tell us about your experiences with sponsoring. how do you sponsor someone?*

I will tell you about my experience in sponsoring Jannik Sinner from Sesto, a city in Italy with 1,500 habitants just 15 km from the border with Austria.

Until he was thirteen years old, this boy played tennis three times a week, while becoming an Italian national ski champion. He realized that practicing skiing was limiting his tennis, because it was slowing him down. He preferred tennis over skiing, because tennis is a game, whereas skiing is a competitive individual sport. In tennis, you play against someone or in doubles. Additionally, a tennis game is about 1.5 h, whereas going down a ski slope takes you 45 s.

He talked to his parents and told them he wanted to dedicate more time to tennis, and have a tennis coach, and that in particular he wanted me, Riccardo Piatti.

I did not have a tennis center at the time, but I took the challenge. I want to meet the kid, to get to know him better to consider sponsoring him. I now realize the importance of a champion's personality in success.

I saw him playing, and I felt the urge to adopt him, in sports speak, that is. I wanted to learn about his personality. So, I arranged with his parents for six to seven training sessions before September of that year and after those, I took him to the tennis camp at the Elba Island. I observed the child, and recognized that, besides physical and muscular capacity, tennis skills and knowledge, he had the right personality to be champion.

Q. *What is the champion personality in tennis?*

He must have a very strong ego. He must be strong.

I was looking at this kid of thirteen years of age, commanding those of eighteen. What a personality! He was strong. Not a kid around older teenagers. Not at all.

So, I decided to sponsor him, and I now train him for free, and created a technical team to prepare him for the tournaments and improve his technique. I also provide him with the right connections and liaisons. The kid exploded last year. As I mentioned, Sinner won at eighteen years of age the Milan "Next Gen" Open, becoming the number 1 of his age category, and number 73 in the world. At eighteen!

I became his sponsor with a signed contract. I am investing in him, and this is a sign of my absolute trust in his future. I am selecting which sponsorships to accept and looking for financial investors that we can use to further his career as much as possible.

I can see a brilliant future for him.

When the right moment comes, and he will have to compete at Grand Slams, I will find the best coaches for him, because it does not necessarily have to be me. It is always healthy after four or five years of good coaching to listen to other opinions. Improvement is also to change for the betterment of oneself.

My concept is that to achieve the best results, you need other people. I am not a magician or a know-it-all. There are others. Confronting, testing, and comparing yourself with others is always good.

This approach enables me to pull out the best of every individual I mentor or sponsor.

Q. *In your own experience, what are the questions that the mentees should ask you?*

The most important thing a mentee should ask is how to gain experience in his or her profession. There are plenty of books on how to do things, but you cannot buy experience.

Even if I teach young professionals in tennis everything I know, it will be difficult for them to replicate exactly what I do, because the gap is the experience I have, accumulated over decades.

The only beautiful thing you gain becoming old is that you have more experience than the younger people. Everything else that comes with age, I don't like.

Q. *Do you think that single sessions, or mini-mentoring sessions are also effective?*

I work extensively with young coaches and it is part of my character to indicate the path ahead. Many meetings are held virtually. Now with the coronavirus pandemic, I have come forward with many new ideas, some of which come to me by observing my people work. I have 25 coaches and technical experts in tennis, physical performance, physiotherapy and more, thinking about how to transform the tennis coaching business during this pandemic. I want to see what their replies will be.

I do not want my teams to create a copy of a single model of leadership or experience. I'd rather see what it is that they propose. I will lead in shaping those ideas with my experience. In coaching and mentoring, listening is vital.

Q. *Do you think mentoring must be in person, or are on-line, digital mentoring solutions equally effective?*

I can and do implement coaching remotely, even by phone. I call Jannik Sinner, my sponsored teenager to comment on sets of Nadal, Federer or Djokovick, and analyze the details of movements over the phone, right after the games.

Q. *Have you mentored or coached women in tennis?*

I had not coached women, with one exception. In July 2019, right after Wimbledon, I received a call from the manager of Maria Sharapova. She had arrived on location with a laceration on her shoulder.

(*At this point of the interview, Piatti mentions that that very day is Sharapova's 33rd birthday*).

So, we agreed that she would attend the tennis sessions I direct at Elba Island, where I host my fall tennis camps. We had a few sessions, before the Australian Open, and then, for several reasons, inclusive of age, physical condition and others, she decided to retire from the game. We worked together only three months. Destiny had it that we found each other too late.

Those who arrive to be number one know their priorities. She was clear with her priorities.

When Maria entered in the tennis court, she would implement my instructions or guidance, expertly putting into action all that was needed. She knew what was to be done and she did it.

Same thing for preventing injuries during faster movements or different angles. She would implement my advice. She is that kind of person.

Q. That reminds us of top players. For example, do you think Nadal will be a trainer upon retirement?

Why do you think that? I think that the most "normal" of all tennis stars are the top ones. The ones in middle ranks are the difficult ones, who cause problems and make an issue out of everything or out of nothing! The person in rank 100, 200, is full of problems. He or she has a lot of questions and is not efficient.

Those at the top know their priorities and their issues and solve them. I believe the players of high level are very normal and I make a point of mentoring this into Sinner. The importance of simplifying things, because life is simple.

Q. Some people have had a special influence in our careers. Mentors and most particularly sponsors are some of those influential people. Who was influential for you?

For me, it was Vittorio Roiati.

Also, Gianni Clerici (a famous coach of tennis in Italy), when I was young, guided me. He sent me to America, where I worked and learned in Florida and at Southern Methodist University (SMU) in Dallas. At that time, there was a trainer at SMU, Dennis Ralston, who trained Chris Evert. I was there to learn how to coach top talent. That is why I went to the US.

Q. Is age important in mentoring?

I base my mentoring and even my work on experience and knowledge. Hence, a mentor must have these to be able to mentor anyone. She or he would need to have lived intensively, to gain knowledge. So, it is difficult for me to imagine a young fellow could mentor, as he would lack the knowledge.

Q. What do you wish you would have done differently regarding mentoring or sponsoring, as a mentor or as a mentee?

I worked for 17 years with Renzo Furlan and then another 17 years with Ivan Ljubičić, who reached to ATP number 3, and now coaches Roger Federer. I spent many years with single individuals, very long periods.

I also had the opportunity to coach Novak Djokovic, and in his case, I made decisions more with my heart than with my head. When I refer to heart, I mean that I let my feelings prevail. This and other experiences taught me I should decide more with my head than with my heart.

I am 61 years old now, so, If I do not change my decision processes now, when will I do that? Nevertheless, I know myself, and I am a very much driven by my emotions. I am happy as that is my personality.

I invest in personalities with my work. For me, personal relations are important with my mentees.

Q. Did you enjoy any mentoring/sponsoring experience?

The last one!

Always the last one is the best experience.

A Shared Selfie

- **Your favorite role models**: I would pick a few from other sports, and those are Massimo Allegri,[1] Arrigo Sacchi,[2] and Julio Velasco.[3]
- **One word to define your experience with mentoring**: I will use two words: Knowledge and Experience.
- **One word to describe your experience with sponsoring**: Trust.
- **Who you wanted to be your mentor/sponsor, but you never had the chance to ask?** No one in particular, I have always asked what I needed to ask of whomever I considered necessary. I have been very open and looked for advice at all times.

[1]*Massimiliano Allegri (Italia, 1967), Football (soccer) technical director and coach, who led Juventus to win 8 successive Italian National Championships trophies, a record for all clubs in Europe. He was elevated to Italy's Football Hall of Fame.*

[2]*Arrigo Sacchi (Italy, 1950), was named one of the best-ever football (soccer) coaches by Soccer World magazine. Technical Director of Italy's national team sub-champion in the World Cup 1994 and led Milan Football Club (1987–1991) to become one of the best football teams, being nicknamed "the immortals of Sacchi".*

[3]*Julio Velasco (Argentina, 1952), is a sports leader and coach of volleyball, technical director of the Italian Federation of Volley Ball, who became famous for his role as coach of the Italian Volley Ball team 1989–1996, leading the strongest team ever in Europe, with the nickname "the phenomenal generation".*

- **Who would be an ideal mentor for you at this point in your life?** Papa Wojtyla (St. Pope John Paul II).
- **What would you ask this person?** To help me to understand the sense of life.
- **What question would you like this mentor to ask you?** What have you done to give sense to your life?

Afterthought

Tennis produced an unusual first: on 11 July 1900. Charlotte Cooper, a tennis player, became the first woman recorded in the Olympic history books as the winner of an individual event. Maria Angela, who was born in Venezuela, wanted to know if Coach Piatti had any experience coaching Garbiñe Muguruza, a talented Venezuelan-Spanish player in tennis, who was ranked number 1 in 2017 and is currently at ATP rank number 16.

Piatti confirmed he also coached Muguruza in 2019, for ten days, in special dedicated training sessions. It was a positive experience and Muguruza is a great person.

Remarkably, we noticed Riccardo Piatti never used the words "player" or "client" during the interview, preferring to use the names of his mentees or the words "person" or "individual". For example, he remembered not only the birthday of Sharapova, but also her age. He was consciously propelling the human being to stand above the numbers. In the world of top competitive tennis, where numbers and ranks are of the essence, this was nothing short of remarkable.

Human, insightful and humble are the words we would pick to describe this giant of tennis.

Ulrike von Lonski

"*With Sponsoring, the Mentoring concept goes beyond and above.*"

M. A. Capello and E. Sprunt, *Mentoring and Sponsoring*,
https://doi.org/10.1007/978-3-030-59433-6_9

A Glimpse

When applied to an individual, the definition of a "pillar" is "a person regarded as reliably providing essential support for something." That perfectly describes Ulrike von Lonski, who is and has been a pillar of all the organizations and initiatives of which she has been a part. Since May 2019, Ulrike has been the Chief Operating Officer of the World Petroleum Council (WPC), an important organization for the energy sector. Ulrike joined WPC in 2005 as Director of Communication. In her current role, she manages key initiatives of the WPC and organizes its triennial Congresses. At the time of the interview, Ulrike was focused on the 23rd World Petroleum Congress, which will be held in Houston in December 2020.

Ulrike is a multifaceted professional who wears several hats in key organizations. She is a member of the board of the Energy Access Platform of the OPEC Fund for International Development and serves on the IPIECA's Communications Task Force, the SPE HSSE Sustainability Committee, and the Visit Houston Customer Advisory & Innovation Board (CAIB). She is one of the authors of the global Gender Report with Boston Consulting Group (BCG), and the Managing Editor of several industry guides. Ulrike has a passion for the development of young professionals, is closely involved with the WPC Young Professionals initiatives and has played a leading role in the organization of the WPC Youth Forums in Paris, India, Calgary, Rio de Janeiro and St Petersburg.

She studied Business Administration in Germany, and she began her career in 1992 with the Adam Smith Institute in London, developing the Institute's finance and energy portfolios for the Former Soviet Union. Subsequently, she became Director of Energy and Investment Promotion, before moving to a similar role in the International Trade and Exhibitions Group. In addition to working with Russia and the FSU on many high-level industry meetings and senior government conferences, Ulrike has experience working with the Middle East, Asia and Africa.

Of the people we have interviewed, Ulrike is one of the most enthusiastic about sponsoring.

- Studied Business Administration in Cologne and Saarbruecken in Germany.
- Has been the Chief Operating Officer for the World Petroleum Council since May 2019.
- Started her career at the Adam Smith Institute in London in 1992, building the Institute's finance and energy portfolios for the Former Soviet Union.

After a short time, she was promoted to be Director of Energy and Investment Promotion.
- Afterwards, she was Director of the International Trade and Exhibitions Group of the Adams Smith Institute.
- Member of the Executive Board of the second WPC Youth Forum in Paris, 2009.
- Presenter at the third WPC Youth Forum in India, 2010.
- Senior Advisor to the Young Professionals organizing committee of the WPC Youth Forum in Calgary, 2013, the Future Leaders Forum in Rio de Janeiro, 2016, and the Youth Forum in St. Petersburg, 2019.
- Co-Author of the Global Gender report with BCG, "Untapped Reserves", highlighting the status of women in the oil and gas workforce worldwide.
- Board Member of the Energy Access Platform with the OPEC Fund for International Development, IPIECA's Communications Task Force.
- Member of the HSSE Sustainability Committee of the SPE (Society of Petroleum Engineers).
- Recipient of the "Kazenergy" Medal for her contributions to the promotion of international cooperation in the oil and gas sector.

A Personal Snapshot

Ulrike von Lonski has been highly effective in promoting multiculturalism and empowering the younger generations of professionals in oil and gas. In her work with the World Petroleum Council, she has championed important initiatives to enhance the visibility and relevance of emerging professionals in global events about the future of oil and gas and energy in general.

Ulrike's network is one of the broadest and most effective in the energy space. Her insights are valued not only within the WPC, but beyond. Many people seek her advice and benefit from working with her on task forces, organizational boards, and committees.

Our interview with Ulrike revealed surprises and we are delighted to share her insights on mentoring and sponsoring.

The Interview

Maria was vacationing in Rome, while Eve was in California and Ulrike was in London. The geographic and time zone challenges in arranging the interview with Ulrike are typical of the challenges of the global, multicultural energy industry.

Q. Please tell us about your own mentoring experiences.

Thanks for this question, which I think is very interesting. I have participated in many rounds of mentoring, and I want to have a mentoring program in WPC. It is shaped for our Senior Leaders to mentor our Young Professionals. They focus on how young professionals manage their tasks, how they engage with our members, and the process and progress of the working groups that they have selected. Although it was very ad hoc and informal at the beginning, these mentoring arrangements have evolved, while keeping their framework very natural.

Q. What is your mentoring process, Ulrike?

I generally make myself available. Through our work together, I have become a mentor to a number of our young professionals. I let them know that I am happy to discuss their issues in more detail. What follows is a conversation, or a series of talks, in which the person does not just ask about specific elements, but also seeks clarifications on how to manage the matter at stake.

Our WPC people engage through working groups, so it is critical that our young professionals quickly become knowledgeable about how to follow-up with volunteers, and how to engage and communicate with them. It extends to a much more personal level.

You ask yourself, "*What else should they be looking at? What else do they need to do and know?*"

I care about my mentoring process, and many times I find myself thinking about what kinds of practical skills they will need in their processes. They all involve sharing knowledge and information. Many times, this it is a matter of helping them to effectively communicate their objectives. A key area for me is communication and marketing. This includes marketing yourself.

My mentoring focuses on project managing, presenting themselves and their work, and personal development.

Q. *Have you found challenges in cross-cultural mentoring?*

Cultural differences have not been a big factor in my mentoring sessions. When the discussion is about an individual and their next steps, mentoring was more about how to deal with impacts from outside than from within.

Communication is at the heart of it. How do you feel about yourself and your achievements? How do you communicate that to others? Even when it is about the individual, you must focus on how to communicate results. It is not really a cultural issue and I have found it does not vary by geography. One concern that many young women and men seem to share is, How do you best present your achievements without coming across as boastful or conceited?

Many young women in particular shared with me that they are not sure about how to present themselves and their achievements. They are proud of their success and want to talk about it, but are worried they might come across as bossy, pushy and overly ambitious, when all they want is to be is recognized and considered for future opportunities. This is where I can help with some of the communication training that I developed during my career.

Q. *Have you had to change any aspects of your personality?*

I have grown up within several cultures, and that makes me aware of different viewpoints and perspectives. I have learned that making clear points and plain speaking brings best results, which means I often come across as quite strict and strong. It is about conveying points in a very short time. Is not about emotional feedback, but about practical input: facts and actions.

I often tell my mentees to keep emotions and other issues out of those conversations. Recognize how they feel, but focus on the achievements, the potential, the educational level, the way ahead, rather than lead with an emotional argument.

This answer might not always be appealing, as many people are looking for emotional support, but I don't think that is the purpose of mentoring. The ones who are really interested in developing their careers, generally agree that it is more helpful to receive practical advice.

Q. Have you found a recurring topic in your mentoring sessions?

I would say the thing that comes most to me is better communication. "How do I best bring my point across? How can I convince/engage others? How do I get recognized for what I do?" are common questions, and I would say even common concerns. These are definitively the main issues.

Questions like, "How do I stand up when I am being ignored? How do I make myself heard?" are constantly asked in mentoring sessions. I see that it is not just about their industry knowledge or competency. It is how they come across. And that is the same wherever you go.

Q. Now, tell us about your experiences with Sponsoring. Have you been sponsored to advance in your career? How?

You won't believe this, but the very first time I ever heard about "sponsors" and "being sponsored" in a career sense was in the global gender study of women in oil and gas which we did with BCG, and that was only two years ago.

I think that sponsoring is very much an American concept. Once I heard of it, I realized it was essential, and I push hard for it now. Key is to find yourself a (good!) sponsor and you should also look for opportunities where you can sponsor others. Sponsoring goes above and beyond the mentoring concept, as I have realized when I applied this concept. It is about speaking up on behalf of a person and pro-actively putting an individual forward for consideration when you are in a position to do so. It's shining a light on that person and opening doors for them.

I have been sponsored repeatedly by very senior people. I was not aware of it back then, and I did not do anything specific to get sponsored, but I must admit I was very lucky. Very senior leaders recognized what I could do, my potential, and provided introductions and connections that took me further. I never realized I could receive sponsoring from my own initiative, as it was given to me, and I was a passive receiver. But now, I understand it is something that may be sought, pursued.

You should actively look for these kinds of enriching and valuable relationships and create win-win situations. It is definitively something I speak about now. I think people need to create opportunities for others.

I have done that repeatedly in my career when I see someone with great potential to be a real for high-flyer, and I mention them to others. Some of it has been very public. For example, when I write a LinkedIn recommendation, or when mentees are looking for further education and I write letters of recommendation for them. Some of these actions effectively empower people and support them to achieve their goals.

I would like to highlight a specific case of sponsoring and that is with our speakers. We try to can sponsor young women in our big events by appointing them as moderators and by giving them visibility in front of broader audiences. I am still very interested in understanding what would be a good mechanism to implement this idea on a larger scale. We still think we need moderators with extensive experience, but that role is an excellent one to showcase young professionals, especially to help young women move into leadership.

Q. Do you have any recipe?

I would not say it is a recipe, but a recommendation. Everyone does things differently, but if you admire someone, go and talk to them and ask how they got to their position. Let people know you are up for more responsibility, more opportunities and create an informal sponsorship relationship.

Q. Ulrike, what are the questions that you would like to be asked in a mentoring session?

I have not really been involved in many formal mentoring relationships. They happen because I work closely with individuals. I notice something they said and that they do not ask the questions they should.

Particularly in a mentoring relationship, I consider you need to think about where you want to be and what you want to be. A lot of the mentors are senior people without much time. The onus is on the mentee to think about what they want and how best to use that valuable (and scarce!) time. Create a dialogue and other issues may come up in the conversation. It is up to the mentee to be clear about where they want to go and what kind of input they want to receive.

As we mentioned before, critical questions like, "How do I present myself?" and "How do I deal with resentment and pushback?" are often asked in mentoring sessions.

Women often put their head down and focus on their direct work, and unfortunately do not look ahead. They do not see the importance of the non-work-oriented activities in their organization, like networking or even sponsoring, losing sight of what is beyond in their career. They do not look at "How do I move from where I am to a different place." Instead, they continue working and working hard in the same place, convinced someone will notice the tons of good output they produce.

But suddenly, and inevitably, they look around and notice that everybody else has moved on to better roles. They missed out on steps along the way that would have taken them in other directions or upwards. "Nobody told me. Nobody told me that I had to look for mentors, that I needed to promote myself more." This comes up in my conversations. Those questions are not always asked.

I stress the need for a slight modification in the way we communicate. When I hear someone say, "We managed to do achieve this" or "My team was really good...." then, I ask them "who leads the team? Who took the initiative?" If the answer is themselves, I suggest that they say that as, "The team I am leading does this and that." This makes a huge difference, as it is then clear who drove the process.

This can become very frustrating for them, as they are often expecting to be recognized on the basis of the work delivered and did not realize that, they also have to promote themselves more. When someone just talks about "we," there is a need for them to learn to talk about "I."

(***Comment***) *Ulrike tells us there are often issues when younger people have to manage older (male) colleagues or contractors. This is especially the case for younger women, even when they have the grade and qualifications to do so.*

That is when others need training as well, especially around unconscious bias and how to overcome it. Our industry needs more young women. We need to give them the opportunities and the support. We still only have a handful of women as it is mainly a male environment, and they need all the help we can give them in order to change that balance. If there are only a few women in the operational areas and in leadership, their successes and their failures become not just their own, but representative for all women in the industry. Judgmental phrases such as, "Women can't do this" or "Women aren't good at that" easily pop up, but if there were ten women doing the job and three fail but seven succeed, then it doesn't become such an issue anymore.

Ulrike and the authors agree there is a need to put together a list, and most importantly, to communicate globally about networking opportunities for women and for public mentoring. Senior women (and men!) should make time to mentor younger ones. We need to eliminate "manels" (panels with only men as panelists) at industry events and instead use them to profile successful women in oil and gas and show young people that there are also female role models out there.

Maria mentioned the recent trend in several countries in the Middle East, where a nationalism trend is taking place, very vocally, as published in the newspapers. The rejection of the different. The advocacy for a not-plural "us", magnified by the handling of the pandemic experience. Very young people come up to here to work and are not ready to experience this kind of pushback.

A Shared Selfie

- **Your favorite role models:** It is a combination of different people. A merged role model, like a puzzle with many pieces. Probably somebody like Michelle Obama comes closest to it.
- **One word to define your experience with Mentoring:** Knowledge sharing.
- **One word to define your experience with Sponsoring:** Empowering.
- **Who you wanted to be your mentor/sponsor, but you never had the chance to ask?** I cannot say. I have been very lucky, in that the people who I go to ask questions have been are very good. Some of the people I have worked with have enriched me. I like the people who have values.
- **Who would be an ideal mentor for you in this moment of your life?** Malala.
- **What would you ask this person?** How do you go from having a strong conviction to turning it into a message that millions and millions now follow?
- **What question would you like this mentor to ask you?** What do you want to do with your life?

Afterthought

Our own thoughts about sponsoring become sharper and more insightful with every interview. Ulrike's take on sponsoring was particularly enlightening. If she was proactively sponsored from the very beginning of her career, she did not realize it until recently. We liked the approach she takes to expose young talent and especially her emphasis on connecting experienced and senior people with younger ones. Isn't that what mentoring and sponsoring is all about? We were inspired. Ulrike shared reflections on pivotal topics with her characteristic insightful perspectives. It was moving to think that just arranging the encounters and networking opportunities for the young to approach experienced people might be the impetus for mentoring and especially for sponsoring. As simple as that. It was a wonderful thought.

Elizabeth Coffey

"*I don't have to be a CEO to coach CEOs.*"

M. A. Capello and E. Sprunt, *Mentoring and Sponsoring*,
https://doi.org/10.1007/978-3-030-59433-6_10

A Glimpse

According to the Cambridge Dictionary, "Spark" is a word that means "a feeling or quality that causes excitement". That is the name selected by Elizabeth Coffey for her company, Spark Leadership. Nothing could better reflect the passion with which she leads her consulting firm which encompasses two operations teams with 10 and 25 associates in 3 different reporting jurisdictions, spanning cultural settings and countries, serving executives in blue-chip private conglomerates as well as in governmental organizations and in academia.

Elizabeth is magnetic. We cannot use another adjective to qualify the appealing way she has to express profound ideas in a direct way, creating a self-reflection in whomever is lucky enough to come across her.

- Leadership Coach to Chief Executives, Vice Chancellors, CEO Successors & Executive Board Directors of FTSE 100 & Global 500 companies.
- Won Pioneering Woman Leader Award 2020 (Economic Times), Top 10 UK companies for Leadership Development 2020 (HR Tech), Top 101 Global Diversity and Inclusion Leaders 2019 (Economic Times), 3 time winner of the Woman Super Achiever 2018-20 (HR Congress), Mentor of the Year 2007 and Women of the Future Awards, 2007.
- Built a profitable business from the start-up phase onwards, attracting world-class professional colleagues from Europe, North America and Asia as Associates; Rainmaker dubbed "a one woman multi-national" by the CEO of a global advertising agency.
- Her clients have included senior leaders at McKinsey, Credit Suisse, Citi (London, NY), University of Liverpool, Petrofac (Sharjah, London), Smith & Nephew (Dubai, Rome), IBM, KKR, Barclays (London, NY, Pune, Singapore, Toyko, Hong Kong, Lithuania), UBS (London, Chicago & Switzerland), Investec, BP (Iraq, London), Shell (London, The Hague), National Grid, PwC (Jordan, London), GSK (London, Philadelphia), Sanofi (Paris), Pret a Manger (London), ADIA (Abu Dhabi), Thenamaris (Athens, London), EY (London), Exxon Mobil (Abu Dhabi), ADNOC Group (Abu Dhabi), TT Electronics (UK), Meggitt plc (UK), Electrocomponents (UK), Atkins (London).
- Coached all female candidates to CEO from the national oil company of Abu Dhabi, ADNOC Group, and all 14 female Ministers in Prime Minister Tony Blair's Cabinet.

- Lead author of "10 Things that Keep CEOs Awake" and co-author of "The Changing Culture of Leadership: Women Leaders' Voices" which sold c 3000 copies in the first year.
- B.Sc. and BA (cum laude), Psychology and English, Wellesley College.

A Personal Snapshot

Maria met Elizabeth in one of the executive offices of the top floor of Kuwait Oil Company corporate headquarters, in Ahmadi, Kuwait. It could not have been otherwise. Elizabeth is always dealing with the top executives of very important companies in terms of revenue, number of employees or strategic standing in business, finance or humanitarian causes. She mentors, tutors, develops, enlightens, coaches, and yes, she literally sparks the career journeys of executives.

At that time, she was coaching one Chief Executive and five of the Deputy Chief Executive Officers of the national oil company of Kuwait, KOC. Maria was asked to pave the way of the coaching sessions, and to invite Coffey to speak at a women's empowerment conference that Maria was organizing as program chair, with the Society of Petroleum Engineers. It was 2012, and it was a pioneering initiative in a moment in time when very few conferences were organized on that topic. Add to that in the Middle East, with the cultural challenges this initiative entailed, it was remarkable. Liz and Maria were both thrilled with the idea of speaking about how and why to empower women in the oil sector. The conversation flow was rapid with the happy frenzied speed of people talking with passion.

They continued collaborating. Liz, as everyone calls her, is more sparkling than ever, and her interview is one of the most essential elements in our book.

The Interview

Liz' interview was an enjoyable conversation. One that ignited, many sparks. She opened our eyes to the dramatic importance of sponsoring and how many more we needed than we had imagined.

***Q.** Tell us about you. What do you do and why?*

I coach senior executives and also advise them on change consulting. I have worked in developing executive leadership for more than 25 years. For the last 20 years, I have worked specifically in change consulting. I wrote a book and have also co-authored books. The first book was about leadership from a female perspective; it was pioneering—the first book in the UK to tell the story of leadership from the female perspective.

In 2002, I was invited by McGraw Hill Business to write a book on Developing Chief Executives. I have also contributed chapters and articles over the years on leadership diversity and inclusion, D&I. I have engaged in large-scale consulting about D&I, including moving black and minority ethnic talent into the top 1% roles in the UK civil service. I led a project developing women and men into to top roles in Citibank's Investment Bank in EMEA and North America, and I designed and delivered a two-year program in ADNOC to progress women into CEO roles, build their pipeline of female leadership talent and make the culture more inclusive at the top with their most senior men.

Q. *Where are you from?*

I am from New York, but have been living in London almost all my career. Maybe I am already a Londoner. I recently spent half of my time in the United Arab Emirates, and in India, where I've done a lot of work for social enterprise. I want to share that I have spearheaded my recent mentoring and development programs in three different continents, in three different faith-based countries with three different economies. Fascinating experiences.

Q. *Please tell us about your own mentoring experiences: have you mentored many people? Do young professionals reach to you to receive mentoring? How does your mentoring process start?*

As a prelude to our conversation, Liz said, "I want to understand what it is that you mean by mentoring and coaching, because in the last 25 years, the concept has evolved."

We realized talking to a specialist was going to be very different. We replied that we envisioned mentoring as a process to guide the individual in her career path, and sponsoring to be when a person with power and/or influence supports the candidacy of an individual for a role. Sponsors consciously enable leadership opportunties for other. We differentiated between mentoring and coaching, explaining to Liz that we considered coaching to be an intensive process. In contrast, we envision mentoring as less structured and possibly even occurring during a causal conversation.

Liz thought it was important to clarify our mutual understanding of these terms and reviewed with us the story of Mentor in Homer's Odyssey. The definition of mentoring comes from Greece and Liz grew up with the idea that Greece was the cradle of Western culture.

In the Greek epic, The Odyssey, Odysseus is going to war, to reclaim Helen of Troy so, Odysseus must leave his son, Telemachus, behind, in Ithaca, where he will continue to be educated. Odysseus chose an older member of the court, Mentor, to guide his son while Odysseus was away at war. This mythical elder guide is the origin of the word "Mentor."

Mentoring is a content-rich experience, a guiding one.

Whereas coaching is envisioned as a Socratic method of questions and answers. Socrates believed you cannot teach anyone anything. Rather, Socrates thought each person has knowledge within and the coach's role is to help bring that out of the individual through challenging questions. Something along the line of hearing back what you are saying. Socrates described this process as being a midwife, helping others to give birth to their own ideas. Purist coaching uses the Socratic method of questioning, challenging and mirroring back to people and is content-free.

Elizabeth explained that meant, "I don't have to be a CEO to coach CEOs."

She added, "I am strong enough in Socratic method such that I know which questions to ask."

Mentoring and coaching use the same methodologies and are not specifically related to non-technical or technical matters.

Elizabeth explained, "I became keen on the methodology, because my father taught us Socratic questioning beginning when I was a kid. He was a Professor of Philosophy, specializing in the Greek philosophers, and he used to teach all of his children during our meals, together. I think was 8 years old, and we already were working on my Socratic questioning."

She added, Looking back, I think I always had this kind of personality that attracts people to come to me to help fix their problems. For me it is a vocational activity. What I realized is that the more I learn about corporate politics and organizations, the more I could add content if needed to a purist coaching approach. My primary client base is CEOs. Those people have very little time. They want you to share content, that would be helpful for them as senior leaders. Many times, they share their views about their problems, which are really significant corporate problems, and they are focused on making their organisations better.

I perform coaching and mentoring at senior levels. Depending what they need.

Q. *Indeed, one of the things that impressed me about you is that you labeled yourself from the beginning as interested and experienced only in very high-level individuals, CEOs and similar. How many people have you coached?*

Over the past 25 years, I would say I've coached thousands of senior leaders, including CEOs. For most senior people, strategic influencing is what they do 60-80% of each day. I have been delivering the course, "Politically Savvy," for about 15 years and have taught that course to tens of thousands of people around the world.

Q. *In your own experience, what are the questions that the mentees should ask you? Or in other words, what questions do you want to be asked in a mentoring session?*

If we define Mentoring as an action intended to provide career advice or guidance, I think it must be content-free advice.

Probably, the question I think is the most critical to ask is "*What is the path to the top of this organization?*" because people make assumptions about how to reach to the top. Those assumptions include, "*I have to work harder*", but that is not the answer. Understanding the path to the top is one of the single most important things you must know to escalate roles.

Q. *But, how can you answer that, if you are not part of the organization?*

People have a lot more information that they think they do. They have 2 and 2, but they have not brought those things together, to be able to understand and make 2 + 2 = 4. I have worked in so many organizations that I know how things work.

If I spend a few months among leaders of an organization, I understand the flows quickly. There is a first important piece to understand the pathways of trust and relationship of how people get promoted. Usually there are multiple pathways up the ladder. You need to focus your activities to advance along the right pathway.

Part of the point of your book is that sponsoring is key to progress. Well, yes, and men understand sponsoring, but women do not get it instinctively in the same way. Women have this fixed idea that things will work for them, if they work hard and do good work. The difference resides in the ingrained perception—based on the gender culture in which you have grown.

I was talking to a guy, who was a very senior advisor at one of the top energy companies. He was talking about how he had seven sponsors. He said: "You need a sponsor in order to progress." All men know that you need a sponsor, but he was clear in that he needed seven. Because through time, organizations evolve. People retire, people move out from organizations, people fall out of favor, and—apparently—in his analysis, seven sponsors was the right number to ensure effective endorsement and promotions. In general, women do not know this intuitively. I asked my female CEOs mentees, "Who are your seven sponsors?" They do not have seven, they have two, maybe three, at most.

Mentoring is very important to improving your understanding of the corporate structure. But having sponsors is essential. In my experience, the majority of women do not think like men on this topic. They just do not.

I cannot stress enough that corporate culture is always evolving. People change, or are moved, transferred. So, you must ask the right questions. If your most senior sponsor moved out, what you are going to do to get promoted, supported in a project at the most senior levels in the conversations that are happening while you are not in the room?

Another important question, "To which projects should you say 'yes' and to which should you say 'no'?"

I thought about the Brexit situation and its consequences, because many of the senior politicians, who brought Brexit to the UK are men.

Yet, none of those men wanted the responsibility to advance Brexit, once it was voted in.

Who took charge? A woman, Theresa May. See the gender difference. Men calculated the risk of the situation and shied away from it, while the woman saw an opportunity to advance and grabbed it. Even if it is an impossible job—which this was. This is the Theresa May story: she saw an opportunity to elevate her position to Prime Minister—the second female PM in the country's history—and she took it. Men did not because they knew this job was a poisoned chalice.

Q. *Maria commented that this triggered her memory of a case of two very senior women in the downstream segment of the Kuwait oil sector with a completely different approach to opportunities and risk. Only one was promoted to Deputy CEO, the one who was more cautious.*

What do you want to comment to us about risk?

I spend a lot of time with senior people considering which roles to take and which not to take. What seems on the surface to be attractive may not be upon deeper reflection. I make them focus on looking at the risk factors in the opportunities as well as the potential upsides of the roles.

Q. *If women have a different perspective, can mentoring help?*

Let me give you one practical example. I had a situation in which a woman, who was a Deputy Vice Chancellor of a UK university, wanted to be the Vice Chancellor of a university. She was on the short list for the VC position at another university. The financial situation at that university was awful, and this was known in the marketplace. In addition, there were other hidden risks in the university's situation. When she began to investigate for due diligence, the university's situation looked worse and worse. I told her that it was not feasible to be successful as Vice Chancellor of that institution. I guided her with a single question: "What are your chances of success?" An informed decision is the best decision you may take.

***Q.** So, do you mean it is sometimes an imperative to leave ambition on the side?*

For sure, there will not be only one opportunity in your life. Part of your job is to create those leadership opportunities for yourself. You do not have to say "yes" to an opportunity just because it is an opportunity. You learn to discern when to say "yes" and when to say "no", and how to identify what could make you fail in the role.

There is more to life than just vision and ambition. Considering a more holistic view is worth the effort. What factors would truly enable your mentees or coachees reach their goals, to have a good life? We need to understand when mentoring highly ambitious people, what they really want in life, the legacy with which they will be happy on their death bed. What is success for them?

I take most of my clients to a list of questions: What has meaning for me in life? What brings me energy and joy? Because all of these people are already successful, they can afford to think about that. They are not just trying to feed their families. They have choices. Is important for them to be fulfilled in their jobs—to be energized, purposeful and happy in their work.

They must have answers to those fundamental questions. It could be family that gives a senior leader her most profound sense of fulfilment. Family is usually very important, but in could be their faith, their side business or other elements.

I tell them that some of these boxes—meaning, energy and joy—are "ticked by" their day job. Others are not. For example: health.

***Q.** Do you think single sessions, or mini-mentoring sessions are also effective?*

An emphatic yes.

Mentoring and Coaching should be effective from the first hour. If it is not, you should not work with that person coach or mentor.

***Q.** Is there a need for chemistry between mentors and mentees?*

It is important to have alignment of values between mentors and mentees. For example, I do not want to mentor or coach the Godfather. I would not work with the current leader of the free world. I would like to be able to contribute to a better world, and it is important to me that my coachee or mentee holds that broad value, too.

Q. *Do you think mentoring must be in person or are the on-line, digital mentoring solutions equally effective?*

I think virtual mentoring works very well for most people. Some people prefer face-to-face. This is part of what I find in the Kuwait companies of the oil sector. For example, I coached 20 individual leaders, but only two would even have a phone call with me. All of them wanted in person contact and sessions. In some countries, the virtual approach is not acceptable. Some people like to have visual cues and stimulation instead of a call.

There are certain cultures/countries where there is high sensitivity to security issues. It is well known that the government or others are recording all telephone or video conversations. For example, I had clients worried about the fact that the US government records all Skype calls. If you had confidential material that you were discussing or viewing, screening by your country's security or intelligence is a real issue.

I once had a case of a leader in the UK, who had to search for electronic monitoring "bugs" in his office. There are many sensitive areas in the finance ecosystem. They have lots of information at their fingertips which could be very useful to others, so it is logical they are targets and surveilled undercover.

Many of my clients are very concerned about industrial espionage, as well.

Q. *Some people have had a special influence on our careers. Mentors and most particularly sponsors are some of those influential people. Please tell us about your own experience as a mentee. Who mentored you in a memorable way? Who sponsored you?*

I worked in a company 25 years ago, in which I was the youngest professional by about 20 years. It was a coaching company. There were only three companies in leadership coaching. I was the first woman, and I the only non-British person. So, I had many people around me who knew more than I did about the business. At the beginning, all of these people were male. Some had been technical directors in engineering companies or HR Directors of FTSE 100 companies. They were knowledgeable about how politics work in organizations; they were fantastic sources of advice.

That population were mentors for me.

I haven't had many sponsors because I have been running my own company for the last if years since very early in my career. I have relied on word of mouth recommendations for my business development at the CEO level. One of the people who sponsored me was a very senior leader at a Board level search business. He has referred masses of work to me and my company. That individual has advanced my career hugely.

Q. *Is gender or age important in mentoring? Females mentoring females, or older people mentoring young?*

I do not have issues with this, myself. I am happy to work with both female and male leaders, and I have male and female clients in most industries. There is a perception of males dominating the leadership of several segments of industry, but even in the most male-dominated cultures, men often revere their mothers. So, I find that men listen to my perspective and advice. They accept coaching by a woman.

Additionally, I must share that I come across as competent and confident—especially when people realize they gain value from our discussion. Working with me, they do not care if I am a woman or a man. But you must demonstrate value very quickly, in the first hour of work These leaders are extremely busy and are extremely intuitive—they don't have time to waste.

Q. *What do you wish you would have done differently with regard to mentoring or sponsoring, as a mentor or as a mentee? as a sponsor or a sponsored professional?*

I would like to be able to figure out a way to scale what I do. I am very conscious that I am limited by how much time I can use.

I am geared towards helping other people and not thinking about helping myself. This is a typical female thing, and it is something that I need to be careful about. If I work too much, it can damage my health.

Q. *Did you enjoy any particular mentoring/sponsoring experience? Why? Tell us more about those joyful occasions!*

I would like to say two things:

First, I can talk to five different people and say the same thing to all of them, but one will be open to that and turn it into magic. On the other end of the spectrum, I have had clients who do not use what I provide to them. I have had a handful of clients who create that magic and that is unbelievable gratifying.

Second, I like taking an almost impossible challenge and making it happen. Basically, I was working with a guy, who was one of two internal candidates for the CEO role. The front runner was older, very experienced—unsurprisingly, he was the preferred candidate at the beginning of the six months. I worked with the younger candidate, on various leadership style issues, and we completely flipped the way the young fellow came across. We flipped the decision of the Board in six months, so my coachee was appointed to the position of CEO of one of the largest companies in the UK because of the work we did. It was an absolute joy to be in that situation.

Q. How do you change the impossible into the possible?

How many leadership coaches does it take to change a lightbulb? One, but the lightbulb really has to want to be changed.

A lot about changing leaders has to do with the leader's determination. You can gauge it in one meeting. You can see if a person is open to change or not. Is that person ready to grab the bull by the horns and make change to himself or herself? It is visible in one meeting.

I always had an instinct about people and what motivates them.

A Shared Selfie

- **Your favorite role model**: My father. An academic, with 10 thousand books. A house of learning, almost our religion. I have taken that as my foundation through life. A fantastic sense of humor.
- **One word to define your experience with mentoring**: Vocation.
- **One word to define your experience with sponsoring**: Essential.
- **Who you wanted to be your mentor/sponsor, but you never had the chance to ask**? Easy one. Socrates!
- **What question would you ask this mentor?** I would ask him everything. We are living in a time when democracy is under threat because technology has the capability to invade our social media and modify the public's vote. I would like to understand how to maintain a sense of ethical core and yet cope with the new ruthless Bond villains.
- **What question would you like this mentor to ask you**? What life do you want to live. What legacy do you want to leave when you are gone.
- **Who would be an ideal mentor for you in this moment of your life**? Malala.
- **What would you ask this person?** How do you go from having a strong conviction to turning it into a message that millions and millions now follow?
- **What question would you like this mentor to ask you**? What do you want to do with your life?

Afterthought

The interview with Elizabeth Coffey opened for us a window into the sponsoring and mentoring of those at the top roles of organizations. We were mesmerized by her insightful way of looking at coaching and understanding that in mentoring and sponsoring you must show value within the very first hour. There is an imperative to show value to these executives in the first exchanges. We realized, this should happen at all levels.

Dr. Ramón Piñango

"Only results lead the way towards continued success".

M. A. Capello and E. Sprunt, *Mentoring and Sponsoring*,
https://doi.org/10.1007/978-3-030-59433-6_11

A Glimpse

Our compilation of insights about sponsoring and mentoring led us to interview a wide range of top executives, from whom we learned and discovered very valuable insights. One fascinating interview was with Dr. Ramón Piñango, who is a trainer, a mentor, and an advisor for executives. As a professor emeritus of what is probably the top leadership organization of Latin-America, IESA, Dr. Piñango's philosophy of teaching is as straightforward as his own style in work and life. In his own words, "Training in business is training for practical uses. Therefore, learning must be centered on shared experiences, analyzed based on conceptual schemes which are fully understood by the participants, from which recommendations for action can be derived."

Dr. Piñango is a prominent thinker and the go-to person for inquiring about the why's and how's of Venezuela political and economic systems. He has been labeled as a synonym of IESA, as his trajectory has been twinned to the evolution of this leadership academic institution. From that platform, Dr. Piñango has done more than write articles or tutor theses, he has in many ways shaped new generations of leaders in Venezuela with a loud resonance and positive impact at the continental scale.

Among other impactful aspects of his career, Ramón Piñango co-edited with Moisés Naim, a renowned global analyst, "The Venezuela Case: An Illusion of Harmony." It is a book that as early as 1984, shared data and analysis which warned about gaps in the societal, governmental and economical fabric of the country which was to face a debacle in terms of deterioration of the economy and quality of life after an abrupt political change in 2008. Publication of five editions of this book highlight Dr. Piñango's influence and well-deserved prestige that made it a privilege for us to talk to him about his perspectives on sponsoring and mentoring.

- Professor Emeritus of IESA - Instituto de Estudios Superiores de Administración, Caracas, Venezuela.
- Professor at IESA's Management and Leadership Center in the areas of organizational behaviour, leadership and resilience, family-owned businesses, and relationships among organizations and their social and cultural context. Management Nonresident Fellow Baker Institute, Rice University, Houston.

- Has been President, Academic Director, Research Director, Member of the Executive Board, Member of the Academic Council, Member of the Directors' Councils of IESA.
- President of the acclaimed IESA Debates Magazine.
- Member of top scientific and entrepreneurial organizations of Venezuela, including the National Council on Scientific and Technological Research (CONICIT); Academic Council of FUNDACIENCIA,
- Author of more than 54 publications and articles, and director and co-editor of "The Venezuela Case: an Illusion of Harmony".
- Advisor of the Polar Foundation.
- Honored as Professor Emeritus by IESA.
- Received the "Andres Bello" medal, the second most important Venezuelan civilian distinction twice in its second and third class.
- Doctor of Education, Harvard University, M.A. in Sociology of Education, Chicago University, and Licentiate in Sociology, Universidad Catolica Andres Bello (Cum Laude).

A Personal Snapshot

We did not know Dr. Piñango personally. By coincidence, when Maria Angela told Olga Bravo (a Venezuelan leadership consultant and evangelist of resilience) about her desire to reach out to Dr. Piñango, she discovered Olga knew him very well and was glad to make the connection.

The interview was conducted remotely with Dr. Piñango in Caracas, Eve in Austin and Maria Angela in Rome. Maria Angela was on vacation in a lovely setting overlooking "Campo de' Fiori" where on February 17, 1600, the philosopher Giordano Bruno, who is regarded as a martyr to freedom of thought, was burnt alive for heresy. The setting was an interesting coincidence, as we later reflected on the freedom of thought of Ramón Piñango, one of the few sharp and pungent Venezuelan voices who has maintained clarity and consistency through decades, while gaining well-deserved respect.

Freedom of thought definitively led Giordano to his unfortunate end, but it was the characteristic we valued the most about Ramón Piñango. We gained some powerful insights about mentoring and sponsoring from our conversations with him that we are pleased to share.

Coincidences are good.

The Interview

Q. Please tell us about your own mentoring experiences.

Thanks for this question. When you two reached out to me for an interview about sponsoring and mentoring, I started to think about whom I had mentored or sponsored. It gave me an opportunity to reflect about that, to think, *"Whom have I mentored?"*

I highly doubt anyone who engages in mentoring does so with a planned approach. In hindsight, one realizes that they have been a mentor of several people, including many whom Maria Angela knows well.

Mentoring happens because you discover people who have the potential to do interesting things. Take Olga Bravo, for example. I have been a mentor for Olga. I was her professor and realized she had an interesting profile, was an interesting professional, and was motivated to register in IESA. Olga became a Professor at IESA.

I have also been a mentor in my consulting services. Moreover, I want to stress mentoring is a two-way relationship. It is a liaison between the mentor and the mentee, just like coaching. For mentoring to be deep and enriching, the mentor must also receive guidance from the mentee. Olga has given me feedback about what we do together.

You also realize you have also been a supporter of the development of people who are at different levels, from secretaries to CEOs, and inside your organization as well as in external organizations. Mentoring is a two-way relationship in which an important component is respect.

One must respect the other. The mentee must feel and be respected. What does this mean? The mentee is listened to and considered. It is not a situation in which the mentor is above the mentee. If I am interested in the opinions of the mentee, then, I will genuinely offer my point of view.

Often mentees are part of your organization and if the organization is mid-size or larger, you generally end up establishing alliances with people at your level or in higher power roles to promote your mentees. This is the realm of sponsoring.

You also probably have helped people with whom you do not work, but whose projects or ideas interest you deeply. In my case, I have mentored people working in social projects, even investing my personal money with costs about which I prefer not to talk.

So, many times you take your risks in mentoring. I think that mentoring is co-mentoring. It is a symbiosis and not a one-way, unidirectional relationship.

Q. *Now, tell us about your sponsoring experiences. Have you been sponsored to advance in your career? How?*

Oh yes, I definitively had sponsors during my career.

I would like to highlight that I have sponsored some individuals, who in turn sponsored me. For example, I have had people. who sponsored and supported my selection for study abroad. Later after my return, I was able to support them. You see? Again, sponsoring is a two-way process, just like mentoring.

These are processes of communication and influence which are bi-directional. If one is not open to this permeable and two-way process, the process lacks depth.

I have had mentors whom I left behind, because the quality of mentoring I was receiving was null. In the 1970's, I was contacted by a Harvard Professor, who was consulting for a project in the Venezuelan area called Guayana. It was a MIT-Harvard joint consultancy, and this professor wanted to offer me a scholarship to study at Harvard, but at the time I wanted to gain experience, so I explained my plan was to work for four more years before pursuing any advanced studies.

After a few years working in the Ministry of Education, I realized I wanted and needed to go for a PhD. It was April and the application deadline at Harvard was March 31. I called that professor and he told me, "*Fill-in your paperwork, I will take care of it.*" He arranged all the necessary internal paperwork and approvals from the admissions committee.

I was admitted.

That person was one of my most important sponsors. For me, it was a decisive moment and a step-change in my work and my life.

I was sponsored to enter at IESA, the Venezuelan Institute for Superior Studies of Administration, by Moises Naim.

Q. *Really? I thought you met him at IESA at the time of the project, "The Venezuela Case".*

Yes, the same Moises Naim of the book, the same Moises Naim of the Naim Effect. I must admit it was not easy for him to convince me to come to IESA. When I returned to Venezuela after Harvard, I worked for two years at the CENDES, an organization associated with the Universidad Central de Venezuela, the most important university of the country. After two years at IESA, I wanted to do more research, so I accepted a job offer with IDEA, to work in research on education topics, which was my passion! Then IESA got a new president, Dr. Henry Gomez, who with Moises Naim persuaded me to stay, with the agreement that I could conduct research. I was made Director of Research of the IESA, beginning a tenure in this organization of which I am currently an Emeritus Professor. It has been wonderful.

This sponsoring effort turned into a very rewarding liaison in which we worked as a team, I as Director of Research, Moises as Academic Director, and Henry as President.

Q. And tell us, have you been a sponsor of anyone?

Yes, I have had several students who later became professors of IESA. The guys I call "IESA material".

Q. In your own experience, what are the questions mentees should ask you?

There are no specific questions. I like the question, "What do you think about ….?"

You need to establish a dialogue, a conversation. This requires the highest level of respect for one another. If there is no respect, there is no possibility of effective mentoring or sponsoring. You must be genuinely be interested in the other person.

Q. And what kinds of questions should the mentee ask you?

Something along the line of, "*I think this… What is your opinion*?"

With Olga, for example, I have been unable to reach a colloquial and direct relationship. She addresses me as Dr. Piñango or Dr. Ramón never simply "Ramón" using in Spanish the "tú", instead of the formal "usted," if you know what I mean.

If a mentor is sincere, he or she must place his or herself at the same level as the mentee, to seek a fruitful exchange of perspectives. As a mentor you help the mentee to see things perhaps with a different perspective from a different angle.

Q. Your answer seems to indicate that the mentee must step up and help build the mentoring relationship?

As a mentee you must be clear about you want to accomplish. The mentor will support you in gaining clarity or assessing other ways to reach the goal, but the mentee must be an important part of this.

Q. So, the mentors should have an approving role?

Not at all. The mentor's role would not even be that of a "questioner."

Q. In your opinion mentors do not have to ask questions?

Well Maria Angela, the trick resides in what sort of questions they ask. Not questions or remarks like "*What do you want to do?*" or "*What you just said is wrong.*" I am more in favor of asking the mentees questions like "What *if....*" or "*What about....*"

Q. *Do you think single sessions, or mini-mentoring sessions are also effective?*

Between experts, yes, they can be effective. A mini-session on career-related matters would be less effective. Career decisions require longer with multiple conversations and mentoring sessions. Mini-mentoring sessions, like an over-coffee conversation, would be for specific issues only.

Q. *Do you believe mentoring must be face-to-face or are online, digital mentoring solutions equally effective?*

Nowadays, it is very difficult to say, "*I will not address any significant matter over the phone. It will have to be in person.*" That is simply impossible. Many times, the mentor is a sounding board and you want to see how your arguments are received and perceived by your mentor. Facial expression and the tone of the voice are important. In person, those are readily accessible, but to imagine we can handle all our mentoring liaisons in person is unthinkable and impractical.

Q. *Some people have had a special influence in our careers. Mentors and most particularly sponsors are some of those influential people. Please tell us about your own experience as a mentee. Did anyone mentor you in a memorable way?*

I must mention Moises Naim. It is an interesting case. Moises is 10 years younger than me. Nevertheless, he had ample confidence in me in an evolving polyhedral relationship over many years. When I think about our interaction, I must admit it is a co-mentoring relationship that is not structured or planned.

When I write an article or I am interviewed, I love to seek the feedback of others. People ask me, "*Have you entertained the idea of including this or that?*" This person often would send me her or his own writings to seek my opinion. It is an exchange that can be very interesting.

Through time, after so many feedback loops, one learns who to contact and for what. "*That is a questioning fellow. What will he think about this?*" or "*This lady is not going to provide any challenge for me to think about.*" So, if I need valuable feedback, which will enhance my article, I can figure out who to contact.

In my case, I enjoy my liaisons with people who provide critique. I do not like those who praise, who do not provide any usable feedback but instead a perennial, "*All you do is so good*" or similar reply. If you praise or demote other people's work all the time, your critique is devalued, as it brings no benefit.

I am not in search of appreciation. I am searching for those who may help me in becoming better in doing my things in a better way.

***Q.** Is gender or age important in mentoring? And in sponsoring?*

No. I do not think gender or age is important for mentoring or sponsoring.

***Q.** Has the "#MeToo" movement made it harder for women to find men in power roles who are willing to sponsor them?*

Not where I live and work. I followed that movement with a journalistic interest, but here, in Venezuela, I have not noticed that problem at all.

What I do feel, and this is a personal comment, is something important for me in mentoring that I want to share. I am and I believe I am perceived as a strong character, a bold and strong person. I express my opinions in a direct and strong manner. So, when I must liaise with people with a soft or shy character, I am inhibited, because I feel like I am over running them. I prefer to talk with people who are not shy about telling me, "*I think you are completely wrong.*"

In human relations, you have an intuition if the person is affected or establishing barriers. I feel it. I become embarrassed and start calculating and thinking how to improve the connection. And that inhibits me. It is a problem.

***Q.** Is that the power of communication?*

Exactly! I am quite good at being sarcastic and that can easily hurt people. I love it when I make a sarcastic remark, and someone replies sarcastically. You build a common ground. If not, the mentoring is transformed into a psychologic therapy, doctor-patient relationship.

***Q.** These processes take place within a cultural frame in which there are values, customs and very similar ways to see things, where the sense of humor is quite equal. People tend to choose, to hang-out with similar people, right?*

Of course! It is the pack effect. The flock, the herd, the tribe. In which all think alike. This happens especially on boards. When there is a dissident voice, that person is uncomfortable, weird for the herd.

My world has been the academy, but I have worked extensively in consulting for corporations. I can avail that questioning is not welcomed in the corporations, as it is in the academic realm. It is just not well perceived.

When dealing with corporations, one has to tip-toe around. The prevailing feeling is, *"Do not rock the boat! Do not rock the boat! We may sink!"* That it may be fatal.

Unanimous decisions and lack of dissent are extremely dangerous. Enron is a classic example in which the whole board was cooked in their own, common hot oil, taking the whole corporation to disaster.

***Q.** What do you wish you would have done differently with regard to mentoring or sponsoring?*

I think I have not thanked enough those who mentored or sponsored me. Many times, one forgets the enormous number of people who have supported our careers.

One remembers easily those who have blocked our careers or promotions, but rarely those who have supported us. I have not been thankful enough.

***Q.** Did you enjoy any mentoring experience? One can have memorable mentoring experiences that go back to childhood years. (Maria Angela as an example tells an anecdote of her childhood in her birth town in Venezuela).*

I was born in "Los Rosales, Caracas, in what used to be the El Valle road that now is known as the Nueva Granada Avenue. I was born at home. In the 1940's, there still were are some like me, who were born at home.

I grew up in a family with many members and in which traditions were respected.

Maria Angela, you will not believe this, because I know you just told me you were born in Altagracia de Orituco, and perhaps you will think I am making it up, but in my parents' house, my father had a person whom he trusted, who was from Altagracia de Orituco, Venezuela. Her name was Natalia, and we called her Talita.

Talita started to work with my grandmother as a maid, but due to her characteristics, she was naturally and gradually integrated into our family. When my mother got married, she came to work for my mom and then naturally, for us. Talita would take her vacation days to go to Altagracia and I was her favorite person, the kid she would happily educate and nag, when necessary.

When I was in second or third grade my parents hired a personal tutor, Ms. Carmen Felicia. to teach me math and Spanish and all the important things for school. She would give me assignments, like sentences to be completed. The cat …meows; the dog …barks; the chicken …cackles; …and things like that.

One day, we were reading about something and I was stumped by the word "agonizing". Ms. Carmen Felicia asked me, *"You don't know what 'agonizing' means? Ramón, you have never seen a person die?"*

Puzzled and without any clue, I answered, "*No, Ms. Carmen Felicia….*"

Then, Talita, who had discreetly followed the lesson from afar, came to the living room abruptly, interrupting my tutor, *"Ms. Carmen Felicia, please excuse me, but don't you know that when people are about to die, one never visits with children?"*

Ms. Carmen was muted by that. But Talita was right! The support I received from Talita was unforgettable for me. My first sponsoring, real sponsoring by Talita, who was illiterate.

As I laugh about my mistakes, I recollect all the bad things that happen to me.

Q. *Why did you study sociology?*

My father asked me what I wanted to study. He was extremely respectful. I explained I wanted to study sociology. He countered asking if I could make a living out of that career, because he thought it was a weird occupation, but he supported me up in all my studies.

A Shared Selfie

- **Your favorite role model**: None. I do not have any.
- **One word to define your experience with mentoring**: Communication.
- **One word to define your experience with sponsoring**: Support.
- **Who would be an ideal mentor for you in this moment of your life?** Don Quixote.
- **What would you ask this person?** Am I on a good track? Because Don Quixote is full of wisdom.
- **What question would you like this mentor to ask you?** For that, we will need another interview.

Afterthought

Dr. Piñango, has a vocal presence on social media, and I enjoy following his twitter account in which he declares his almost unfiltered thoughts and where his political standards are evident. We want to share three quotes from him, taken from his Twitter account, which we particularly liked.

For me, education is to bring the individual value out, extract the potential, to challenge and provoke, so that he or she may develop and grow.

The perception of the facts, more than the facts, is what determines action.

Integrity is the foundation of any personal or institutional leadership.

Maestra Maria Guinand

"Mentoring through music has been my main goal in life."

M. A. Capello and E. Sprunt, *Mentoring and Sponsoring*,
https://doi.org/10.1007/978-3-030-59433-6_12

A Glimpse

From the dawn of civilization, music has been delighting listeners. This sublime art form requires skill, preparation and coordination of many accomplished musicians to make the sum of their efforts greater than the individual parts. Lessons from music including perspectives about sponsoring and mentoring provide insights into those practices in dramatically different career paths. We are delighted to include the wisdom of Dr. Maria Guinand, renowned musician, educator and conductor.

Dr. Maria Guinand is a prominent Venezuelan choir and orchestra conductor. Her conduction of concerts and recitals has received not only standing ovations, but also numerous recognitions and awards including Emmy Award nominations, the Helmuth-Rilling Award, the Life Achievement Award of the International Federation for Choral Music of the Bach International Academy and Italy's "Guido D'Arezzo" choir performance award. As one of the key leaders of the Foundation Schola Cantorum, Dr. Guinand has enriched the educational system in Venezuela and benefitted society by enhancing the education of children through music.

More than 18,000 people have been members of the choirs that Dr. Guinand has directed during her career, principally, the Schola Cantorum, the Cantoria Alberto Grau, Scholas Juveniles, and the network of Little Singers in the "*Construir Cantando*" (building by singing) program in Venezuela. She has had broad impact through her preparation of choirs for directors Simon Rattle, Claudio Abbado and Krzyzstof Penderecki. Also, she has recorded 28 CDs of choral music and led fund raising for her activities with multilateral organizations around the world.

- Coordinator of Choral Symphonic projects of El Sistema until 2012.
- Artistic Director of the '*Música para Crecer*', a program sponsored by the Andean Development Corporation (CAF) from 2003 until 2017.
- President, Vice President and Advisor for Latin America of the International Federation for Choral Music (1996–2008) and member of the International Music Council (UNESCO) (2002–2005).
- Coordinator of the Master Program in Music, University Simón Bolívar and member of the Superior Council, University Simón Bolívar (Venezuela, 1996–2000).
- Numerous awards in choral music: 'Kulturpreis' award (1998), 'Robert Edler Preis für Chormusik' award (2000), the 2009 Helmuth Rilling Award and the Life Achievement Award of the International Federation for Choral Music (FIMC 2019).

- Honorary Doctorate degree from the Metropolitan University, Caracas, Venezuela (2020).
- More than 39 international musical performance tours, 56 presentations as conductor, special guest conductor and jury participant at international competitions, and 28 CDs.
- With the Cantoría Alberto Grau, she won six prizes in the choral competitions of Neuchatel and Arezzo (1989) and with the Orfeón Universitario Simón Bolívar three gold medals in the Choral Olympics, Linz 2000.
- Directed concerts and choirs at the Liederhalle (Stuttgart), Barbican (London), Teatro Colón (Buenos Aires), Sao Paulo Room (Brazil), Disney Hall (Los Angeles), Lincoln Center, Alice Tully Hall and Carnegie Hall (New York), Palau de la Musica (Barcelona).
- An accomplished educator, she has focused on Venezuela, but taught extensively all over the world. She has trained other musical directors and is a Master of Music.
- Bachelor and Master of Arts in Music, Bristol University (UK).

A Personal Snapshot

Our interview enabled us to hear her own account of collaborators, supporters, sponsors and endorsers. She rarely said, "I." She speaks in plural, praising others. She is humble, but her numerous awards provide proof of her tenacity and success through her more than four decades of enriching musical teaching around the world.

The Interview

Q. Please tell us about your own mentoring and sponsoring experiences.

The mentoring I received was very important in my career. My first mentor is now my husband, Alberto Grau. When I was eight years old, he was my first teacher of music. He put my hands on the piano. I have stayed with him all my life, beginning, as his pupil for a very long period. Then we had our own separate personal lives and different marriages, but we got together again. My husband, Alberto, has guided my understanding of music and life, especially on how to relate both. He has provided personal, family and musical mentoring with the full range of subjects mixed together throughout our lives.

But let me go back in time, because I was mentored from my early childhood. I was blessed to grow in a happy and united family with seven siblings, and surrounded by my four grandparents, and many aunties and uncles. All these people had different pockets of influence in my life.

I went to a nuns' school, the San José de Tarbes, in Caracas, Venezuela. Contrary to what many people may feel, I remember those women as pious and strict, but also sensible. Many of them served as missionaries in Brazil. Those women had a lot of social sensitivity and planted in me a sense of compassion for humanity that was very important for the rest of my life.

I recognized my vocation when I was very young. I knew from early times that I loved education. There was a literacy program in the favelas led by the government. I was 13 years old when I was asked if I would like to help in that program as a component of my syllabi at high school, and of course, I accepted! So, I would go on Saturdays to literacy classes, teaching old people how to read and write, I became sort of a young companion for them. I learned a lot.

When the time arrived to initiate my university studies, I selected physics and mathematics, but then realized my true passion was music.

My formal education in music had a wide range of mentors.

At the early stages of my career, there were very good masters of music in Venezuela, human beings of wonderful sensibility, who understood the major scope of the world of music and the very language of music. I could mention many, but right now I remember Gonzalo Castellanos, Alvaro Fernaud, Ana Mercedes Rugeles, Cristina Pereira and Angel Sauce. They were human beings with a special sensibility. They mastered music not only in harmony and solfeggio (a voice exercise in which runs and scales are sung to the same syllable or syllables), but in the language of music, transmitting to me their vision. This shaped my early formative years.

Then, I had the joy of studying in England with two very important mentors for me. One was my piano teacher, Kenneth Mobbs, and my professors, Raymond Warren and Whyndham Thomas. They mentored me when I was only 18 years old and had just come from Venezuela. I was the only Venezuelan at Bristol University. They literally opened their hearts to me, supporting and sustaining me through the years I stayed in Bristol. They gave me hope and pushed me. They pressed me a lot and raised my level and understanding.

Although I studied performance at Bristol, I always knew that I preferred to dedicate my life to mentoring and teaching of music. Mentoring through music has been my main goal in life. It has given me a lot of satisfaction. There are not many books written about how to direct choirs.

Later in my professional life, I worked many years in "El Sistema," the framework or the juvenile orchestras and choirs in Venezuela with the mastermind and director of it, Maestro Jose Antonio Abreu. Along with Alberto Grau, my husband, we did all the choral projects of "El Sistema", together with our organization, the Fundación Schola Cantorum de Venezuela. I witnessed in first person the big dream of this extraordinary man, Abreu, to shape a system for all the country, to transform poverty with the power of music, a project that everyone thought was impossible, even crazy. But here we are, El Sistema became an on-going and vibrant program program as well as our Fundación Schola Cantorum.

Abreu taught me about the reality in which we were living, including about the politicians in my country and how we could gain sponsorship and funding for the projects. In short, how to work with the public sector. This was not trivial, given how chaotic our country was and is. He taught me how to create opportunities during chaos. That was for me very strengthening.

Later, when I was in my 40s and 50s, I had the opportunity to opportunity to direct choirs and orchestras. Helmut Rilling, gave me confidence as a woman to conduct not only what I was used to doing, my 'a Capella' choirs, but also to be in front of an orchestra, leading choral symphonic works.

Some people think that a choral director is a person who just goes to an auditorium, stands on a podium and directs from a score. It is not like that. A computer could do that. Music is not reading from a score. It is shaping a group, motivating them, and creating a sense of belonging and commitment to the voice of music.

Not every country is the same. In Venezuela, we must create the enthusiasm about how to sing. You must create this family feeling, this passion, even gusto for singing. "*I don't know how to sing,*" children or adults will tell you. We just don't take that as an excuse. We tell them, "*Come! It is fun, and we will teach you, you will not be alone in this adventure.*"

This was true from the founding of the now 53 years old old Schola Cantorum de Venezuela, with which we have toured the planet. We have directed many choirs and shaped the Schola through the years, but this special bond and feeling remains. Institutions are a big family. We remember not only what we sing, but what we love and share, who was my friend and peer in the choir, which experiences did we have. They see in us their parents in music.

I think mentoring is central to all that we do in music. When I have a problem, I must think how do I address this? I ask for other points of view from many mentors. In Venezuela, we must fight to the end for one idea. I feel like Don Quixote. In Venezuela, we always have big waves like a sea in a tempest.

We did it because we were determined to do it. I remember the first major currency devaluation. We had planned for an Asia tour that due to the devaluation would require twice the budget. People were demoralized, but I said, "We are going." I redoubled the efforts to find new and additional funding, which we obtained, enabling us to go.

All our life was like that. Another example was when we were rehearsing in downtown Caracas and we were caught in one of the huge demonstrations called "El Caracazo" (the wave of protests, riots, looting, shootings and massacres that began on February 27,1989 in Venezuela's capital, Caracas, and the surrounding towns). I was rehearsing a women's choir and they informed me, "the city was in flames." I had to make plans to send each person safely to their home.

(***Comment***) Maria Angela said, "*The lessons you share directing choirs are so powerful. I was one of the mothers who accompanied children to a La Schola presentation at "La Casona," the equivalent of the White House in Venezuela, for a Christmas concert. We waited hours for Chavez' wife, Marisabel. When she came out, one woman complained the snacks were expired, and the choir organizers were jeopardizing the health of attendees. You managed the whole situation so graciously and the concert was wonderful. You indirectly mentored me about how power works and how powerful people relate to each other. You are right, they have a special way of behaving and approaching other people, and this has helped me in Kuwait.*

It has been a constant in my professional life. An important skill I have developed is how to keep together groups in times of crisis. That is done with enthusiasm and with love. With voluntary work on an interesting and important challenge, we create bonds and networks.

Now, we are in the middle of the coronavirus pandemic. I am in the States leading choir rehearsals and managing all the logistics and administrative processes involved. Facing chaos has become my norm.

Q. *Let us now talk about sponsoring. Have you been sponsored?*

For sure! I have been sponsored all my career. When I was studying in Bristol, I received letters from people who later became my mentors, saying, "They needed me," because there were few people with my credentials in my country." After that they gave me the right space to work and I was able to start doing what I needed to do. I needed that push at the beginning.

Alberto saw in me the possibility to develop a long-term choral project and someone who could pick up his legacy and pass it on to others. He is my main mentor.

On my side, I have tried to do the same, especially with choral directors. I try to find jobs for them. I work to have the whole system of choirs recognized and respected. I have done that many times at different levels, from writing an endorsement and recommendation letter to providing counseling. I do it every day.

Q. *Do you want to share with us any names of students you have sponsored?*

I have sponsored many, whom I would like to mention, because I am proud of them. There is Christian Grases, who is now a USC professor. There is Maria Leticia Gonzalez, who is teaching in Boston, Ana Maria Raga, Luimar Arismendi, Diana Cifuentes, Claudia Viviana Espinoza and other Choir Directors in in Venezuela and Colombia. Many more!

Q. *In your own experience, what are the questions that the mentees should ask you? Or in other words, what questions do you want to be asked in a mentoring session?*

I always enjoy the questions that are related to "*How did you get to the point of your career where you are? What roads did you have to travel?*" It has many different possibilities. I like to be asked if I have had any difficulties in my career as a woman. I like that, because I can say, "no".

I never thought that I was diminished or experienced rejection as a woman. It didn't happen. Not with me.

I also like to address the technical aspects of mentoring that relate to music not just as fantasy, but to technical aspects that must be approached daily with lots of discipline.

I like to address the issue of having a long-term project with possible short-term goals, as that enables me to mentor on a step-by-step. I like that.

Q. *Do you think single sessions, or mini-mentoring sessions are also effective?*

Yes, Single sessions and mini-mentoring can be very effective in what you must do and what you don't have to do. Following a good example or saying that is what you don't want to do.

I have had the opportunity to work with major conductors worldwide. One of the best examples of good mentor is Director Simon Rattle, who was indeed a wonderful, well-rounded conductor. His human approach to every single issue was awesome. He took into consideration that he was working in Venezuela, and that he wasn't going to achieve perfection taking the fast approach of a motorway. Instead, he had to take small roads to take us to where he wanted to be. He knew how to do that.

Other conductors, with world-acclaimed status, deluded me with their lousy approach to people.

Q. *Do you think mentoring has to be in person, or are on-line, digital mentoring solutions equally effective?*

Group mentoring needs human contact. I don't think human technology can compete with a person-to-person contact. I can't use a videoconference like this to rehearse choirs. If one speaks, no one else can speak and be understood. I work in music, and on-line solutions will not work for group mentoring. My feedback is the interpretation of music that goes through sound. Choral singing is to create the sound with your neighbor feeling the same emotion.

It is the same for a person-to-person relationship. We need the participation of the other person as well.

I loved the experience currently going on of the Italian balcony spontaneous choirs. People need other people. There is no better demonstration of that closeness than that!

Q. *Is gender or age important in mentoring? Females mentoring females, or older people mentoring young?*

Gender is not important, but age is. Age is experience. You can have many professors teaching you technical things, but few mentors. The mentor will tell you something else. The mentors are those with experience, who can lead you in a different and meaningful way.

Q. *What do you wish you would have done differently regarding mentoring as a mentor or as a mentee?*

If I had to live again, I would be thankful to have half of what I have had in my life.

I have been blessed and I have been aware enough that I could see that and could grasp everything that was given to me.

As a mentor, I would have liked in some cases to have had more experience. I started mentoring very young and didn't have enough patience, compassion and time.

I don't think mentoring should go on forever with the same mentee. There comes a point, when you must let them go. Some people want to be attached, but don't seem to solve their problems. Tell them, "*Time is not eternal. My life has an end. I have so much time for you. Listen, learn and take what is useful for you, but then I have to move on to someone else.*"

Q. Did you enjoy any mentoring experience? Why? Tell us more about those joyful occasions!

A step change happened to me in my approach to mentoring when I was preparing my women's choir to go to a major competition in Italy. I was very proud of my preparation work. When I considered we were ready, I invited my maestro and mentor, Alberto Grau to come to my rehearsal and told him I thought we would win first prize. He started asking questions related to the specifics of interpretation, and I realized there were so many points I hadn't seen and things I didn't know. I was in tears at the end of the rehearsal and realized I had many miles to go to be as good as I thought I was.

A Shared Selfie

- **Your favorite role model**: I admire many people for their achievements in life, but I don't really have a role model.
- **One word to define your experience with mentoring**: Responsibility.
- **One word to define your experience with sponsoring**: Optimism.
- **Who would be an ideal mentor for you in this moment of your life?** Johann Sebastian Bach
- **What would you like to ask him?** How does your mind work? How can you create in such an easy way these perfect works of art?
- **What do you think he would ask you?** What are the sopranos so squeaky? Why aren't they singing better?

Afterthought

Some people have an inborn and distinctive ability as human beings, to stretch their comfort zone to serve others, for the good of society. Maria Guinand is one of those people, and we are glad to include her insights on mentoring and sponsoring in our compilation. We realized that mentors and sponsors are needed in every profession.

After we ended the interview, Maria shared with us that she thought mentoring and sponsoring in music were very different than in business, which is something very tangible and measurable. She was glad that her work was spiritual, because music is intangible.

She told us: "*My only worry is how can this sound better? How can I reach perfection in this passage of the cantata?*" And with this, the interview ended, literally, on a good note!

Dr. Estella Atekwana

"I see mentoring as my responsibility to invest in the younger generation."

M. A. Capello and E. Sprunt, *Mentoring and Sponsoring*,
https://doi.org/10.1007/978-3-030-59433-6_13

A Glimpse

Volunteering in women's networks opens many possibilities for women. Maria Angela and Eve belong to the women's networks of several professional societies. One of those is the Women Network Committee (WNC) of the Society of Exploration Geophysicists (SEG), which was launched in 2011. This pioneering initiative was very much needed for societies related to STEM careers. Eve was the first chair of SEG's WNC and Maria Angela has chaired WNC twice. Through this committee, they have met awesome women leaders. One of those who shines with great brilliance is Dr. Estella Atekwana, Dean of the College of Earth, Ocean, and Environment, at the University of Delaware.

Dr. Atekwana is a geophysicist who has had a very successful career in academia. A leader by nature, Dr. Atekwana has emphasized mentoring and sponsoring of minorities in all the organizations where she has worked and in local communities. Currently, Estella is the Dean of the College of Earth, Ocean and Environment at the University of Delaware, as well as an adjunct professor at both the University of Waterloo and the Missouri University of Science and Technology. She pioneered biogeophysics as a sub-discipline in geophysics which combines environmental microbiology, geochemistry, geomicrobiology and geophysics. She has served as the lead principal investigator (PI) for large, interdisciplinary research projects funded by several federal agencies and industry.

- Dean of the College of Earth, Ocean and Environment at the University of Delaware
- Regents Professor, Sun Chair Professor and Department Head of the Boone Pickens School of Geology at Oklahoma State University
- 265 publications in peer-reviewed journals and conferences with 14 "Best Paper" and "Best Poster" awards
- Keynote speaker at international geoscience conferences
- Enthusiastic promoter of international research and education with emphasis in academic collaboration of USA with Southern Africa, Middle East and China
- Active in 8 professional associations
- Faculty Excellence Award—Missouri University of Science and Technology (2005)
- Master of Science in Geology/Earth Science from Howard University and Ph.D. from Dalhousie University in Geology/Earth Science.

A Personal Snapshot

We know Dr. Estella Atekwana as a very enthusiastic promoter of international research and education. We admire her leadership of study-abroad programs in collaboration with China University of Geoscience in Wuhan, the University of Botswana, the Botswana Geological Survey, Botswana International University of Science & Technology, Damanhour University Egypt, the University of Zambia, Geological Survey Zambia, the Malawi Geological Survey and the Malawi University of Science & Technology.

Through these initiatives, Estella has exposed more than 40 US students to international field research while building capacity in the host institutions. This work has earned her the International Education Faculty Excellence Award at Oklahoma State University (2009), as well as induction as an Honorary Member of both Phi Beta Delta, the Honor Society for International Scholars (2010), and the International Golden Key Honor Society (2008).

Her multiple awards include "Eminent Faculty Award" from Oklahoma State University in 2015, Association of Women Geoscientists 2019 Outstanding Educator, Society of Exploration Geophysicists 2016 Outstanding Educator, Society of Exploration Geophysicists 2020 Virtual Near Surface Geophysics Global Lecturer, and elected Fellow of the Geological Society of America.

The Interview

Q. *Please tell us about your own mentoring and sponsoring experiences.*

Mentoring is essential. It is a critical part of anyone's career. The reason why I am motivated to mentor others is because I never really had any kind of structured mentoring as a young professional, and I faced many difficulties that could have been avoided with proper guidance. I don't want other people to go through what I went through, so I am proactive in providing people with the kind of support that is needed for developing a successful career. As a female scientist, other female scientists constantly reach out to me. Even people I do not know well consider me to be a mentor or present scenarios to me, to gain my insights about how to work through challenges they are facing.

Q. Does the University where you are currently working have some sort of formal mentoring program?

As Dean and senior administrator, I believe mentoring of early career faculty (or faculty at any stage of their career for that matter) is important to empower them, and enhance their success and retention in their positions. It also helps them build leadership skills. In my college at the University of Delaware, we have established a formal mentoring program for our faculty. My college has also developed a formal mentoring program between our alumni and the students. This also serves as an excellent way to expose our students to our alumni network.

I also see myself as a mentor and can't count how many whom I oversee or people within my department or in other departments I have mentored. At Oklahoma State University (OSU), graduate students did not fully understand how to navigate the professional world post-graduation. To address this gap, I implemented a professional development mentoring program for Ph.D. students to expose them to multiple career pathways in the geosciences. All the students who went through the mentoring program have graduated and all of them successfully secured positions in academia or industry.

Q. What kind of mentoring was applied? Group or individual mentoring?

We often engage in group mentoring; however, it was just as effective as one-on-one mentoring, because it was implemented in a very interactive format. It was essentially a professional development program combined with mentoring.

Q. How do students go about the process of applying for this kind of mentoring?

There was no formal application process as it was strictly on a voluntary basis.

I worked with students on topics such as networking, developing a research niche, applying for jobs, contacting potential postdoctoral advisors, writing cover letters, teaching, research and diversity statements, negotiating salaries, presentation and interview skills.

I also provided one-on-one mentoring which included reading cover letters, research proposals, resumes, manuscripts, career development plans; reviewing application packages; and coaching students through dealing with difficult advisors, graduate school applications, etc. I never saw it as work, I saw it as my responsibility to invest in the younger generation to help them succeed. This has always been my passion: investing in people and helping them succeed in life.

When I arrived at the University of Delaware as dean, I implemented mentoring policies and best practices for mentoring. I partnered with the University of Delaware's NSF ADVANCE program (a USA National Science Foundation program to increase the representation and advancement of women in academic science and engineering careers). The goal of the program is to limit bias against women and minorities and help improve retention. Mentoring is required. Every college here has an ADVANCE fellow who is charged with implementing best practices by conducting workshops and other engagement activities. Now, it is established policy in my college that every junior faculty must have a senior mentor.

Before coming to the University of Delaware, I was already aware of the NSF ADVANCE program, with which we paired faculty with junior faculty.

When I look at successful mentorship, I think three elements are necessary: (1) You need a mentor, a coach and a promoter to help you with your career. (2) The mentoring process must be mutual for both the mentor and the mentee. (3) People must understand the value of mentoring.

Just matching a junior faculty to a senior faculty doesn't magically ensure mentoring takes place. All participants must want to do it. I think faculty need to be taught how to mentor and mentees should be taught how to receive mentoring. At Oklahoma State, the NSF ADVANCE program provided training workshops for mentors and mentees, the parameters and the do's and don'ts of mentoring. We organized workshops for the mentors on what they should and shouldn't do.

Successful mentoring requires that the mentee is open to mentoring and that the mentor also understands the value of mentoring. Both parties then have to be proactive in engaging each other. If the engagement is one-sided then such mentorship relationships do not end up being successful.

Q. *Now, tell us about sponsoring. Have you been sponsored to advance in your career?*

I have been sponsored in that I have had someone recognize my impact and subsequently nominate me for a promotion or a role, an assignment, or an award. I recognize these specific actions as sponsoring.

While at Oklahoma State, a junior faculty member believed my accomplishments deserved recognition and nominated me for Regents Professor. My SEG and AWG outstanding educator awards have all been initiated by nominations from my former students. Finally, my election as a Fellow of the Geological Society of America was sponsored by a senior colleague.

When it came to moving into administration, however, I faced some headwinds. When I expressed an interest in transitioning into administration, I was told not to risk killing my research career. We sometimes limit good leaders by preventing them from pursuing their career goals. I believe it is up to the individual to determine what they want to do with their careers. I am now a dean because people in my network brought this opportunity to my attention. They recognized leadership strengths in me and felt that the opportunity matched my ability. I try to be mindful to do the same for others, and have nominated other colleagues for leadership positions.

Q. *In your own experience, what are the questions that the mentees should ask you?*

I like people to ask me questions on how I navigated my personal life, work-life balance, for example, or family and career balance.

Q. *How would you describe the mentoring you received?*

Much of my mentoring has been informal. With students and colleagues, it has been "let's go out to lunch or have coffee" or "come into my office and let's talk." I think this has been much more effective, because in these informal environments I am able to speak from the heart.

Q. *Is mentoring on-line equally effective as mentoring in person?*

For me, digital platforms are equally acceptable and effective.

At University of Delaware we set up a mentoring program between alumni and students by pairing students with alumni and they meet virtually. I have had some occasions when alumni come back to campus and have taken the opportunity to organize meetings so that they get to meet their mentees in person. Given my international network, I continue to develop mentorship relationships with people from different parts of the world.

Q. *Some people have a special influence in our careers. Mentors and most particularly sponsors are some of those influential people. Please tell us about your own experience as mentee. Who mentored you in a memorable way? Who sponsored you?*

Perhaps two people had that kind of influence on my career, but not because there was any kind of structured mentoring. It started very, very early on in my career.

After graduating, my first job was as a faculty member in a near-surface geology and hydrogeology department. I felt like a fish out of water. Not only was I the first female faculty to be hired in that department but also the first non-white faculty hire. This was an extremely difficult environment to be in. In those days, we did not recognize the value of mentorship, so most of my colleagues just left me alone. I guess, they didn't know what to do with me. Nonetheless, during my interview, I had the opportunity to go to dinner with the Vice President for Research at the time who was African American. He gave me the best advice I have ever received in my career. He told me, "If you are to survive academia, you must publish or perish, and you must develop a research niche." This proved to be very valuable career advice.

However, at the time, my research interests (deep crustal geophysics) were completely misaligned with my department's focus (environmental geology, hydrogeology, and geophysics). I struggled to find students to work with me as none of them were interested in deep crustal geophysics. But I kept in mind that I had to publish or perish. I adapted my research program to include near-surface geology in order to accommodate the interests of my students and to become better aligned with my department.

A senior geophysics colleague in the department helped me make this transition by inviting me to participate in his near surface geophysics field research programs. It was during one of these trips, working on delineating a hydrocarbon plume, that we made the discovery that hydrocarbon contaminated sites undergoing biodegradation are conductive, not resistive. This study helped to pioneer the field of biogeophysics. This colleague also invited me to attend SAGEEP (Symposium on the Application of Geophysics to Environmental and Engineering Problems). It was at SAGEEP that I was exposed to a whole different world of geophysics. I took a short course on environmental geophysics and learned how I could adapt my geophysics knowledge to solving environmental problems. I credit some of what I have been able to accomplish today to this colleague of mine, who informally mentored me without even realizing it! In short, we impact the lives of people in so many ways without realizing it.

The second mentor was a connection I made while on sabbatical at the University of Botswana early in my career. I chose to spend my sabbatical in Botswana after a former Botswanan student of mine urged me to go there so that I could be a role model for the female students because the university had no female faculty. I was intrigued and sent a letter to the department chair asking if they would host me for a sabbatical. To my surprise, the department chair responded and I was offered a Visiting Associate Professor position there. They were interested in me because they wanted to develop a geophysics program but had no geophysicist on staff. When I got there, I had a formative conversation with a senior faculty member. He told me, "*You are a female geoscientist from Africa. There are very few of you. You need to think about establishing programs to help build capacity in Africa.*" He volunteered to help me and even suggested quite a few ideas to me. He was always brainstorming with me on how I could help him train students at the university and around Africa. His energy was contagious, and it inspired me to be more proactive in mentoring.

I launched my rift tectonics research program during this sabbatical. Since then, my tectonics research program has expanded from studying the Okavango Rift Zone to include other parts of the East African Rift System. As this project has grown, it has opened up opportunities for me to engage students and junior faculty colleagues in international research. Scores of students from both the US and around Africa have been able to come together to receive training through this program. The colleague of mine that helped me initiate this program passed away several years ago, and, unfortunately, he will never know how much he impacted my career and the lives of others. In life, there are many people who have been what I call destiny helpers. These are people who have a significant impact in helping us achieve our destinies. More often than not, these destiny helpers may never realize or even see the impact they've had.

Q. *Where are you from originally?*

I am originally from Cameroon. I came to the States in 1981 and have lived here since then.

Q. *So, when someone asks you, "Where are you from?" What is your answer?*

I consider myself to be from both countries, Cameroon and the USA.

Q. *Is gender or age important in mentoring and in sponsoring? Are there cultural barriers in mentoring?*

Not necessarily, but there could be. Effective mentoring of students requires that we understand their cultures and their history. I am very careful not to cross certain cultural barriers. I often find that my international students tend to be more comfortable with hierarchical relationships, and are often afraid to question authority. There is sometimes power play that goes on between advisors and their students. For some, this can lead to mental health challenges as they learn to navigate interactions with advisors. American students, on the other hand, are more likely to speak their minds more freely, regardless of the position of the person they're speaking with. This is not to say that they don't also experience mental health issues. However, these issues are more common among international students.

When I mentor international students from developing countries, I try to help guide them beyond their comfort zones. This is because, if they do decide to return to work in their home countries, they will be considered experts. For example, a good number of faculty at African universities have a Master's degree. This is not the case in the US as the majority of our faculty have Ph.Ds. Therefore, I try to get them to learn as much as possible and provide them with the sort of additional experiential learning opportunities that may be limited in their home countries due to lack of instrumentation and technical expertise.

Additionally, it is important for me to provide my students with projects that they are passionate about. I have come to recognize that students excel when they work on projects that excite them. For example, students from minority backgrounds may be more passionate about projects with a strong community engagement component because they like to give back to their communities. As a mentor, I try to be actively flexible and accommodating.

That said, I have some students who are very engaged, and others who aren't—but I don't push them. I prefer for students to recognize the value of mentorship and seeking out mentors. When they are able to see the value in it, they embrace it and want to be mentored more than if I try to force them.

Regarding age, I would say it doesn't matter. Senior people can learn from younger people. I have mentored male, female, white, black, whatever—that should not be a problem. While it's important to be adept at cross-cultural communication, If one understands where the student is coming from, they can be an effective mentor. However, I don't believe in one size fits all.

Q. *Is there anything you wish you would have done differently regarding mentoring or sponsoring?*

As I mentioned, a lot of the mentoring I have received has been informal. As a mentor, I try to instill in people what I love. I see myself not only as a mentor, but also as a life coach encouraging, motivating and promoting my students and helping them excel. I try to nominate them for awards and guide them toward career opportunities.

Q. *Did you enjoy any mentoring/sponsoring experience? Why?*

When I look back and see the accomplishments of my students, those who I pushed or mentored, it amazes me. For example, it gives me great satisfaction to see how far some of my Ph.D. students at Oklahoma State have gone.

A Shared Selfie

- **One word to define your experience with mentoring**: Investing: investing in other people and their success.
- **One word to define your experience with sponsoring**: Recognition.
- **Who you wanted to mentor or sponsor you, but you never had the chance to ask?** I admire people with vision—those who have demonstrated the ability to see into the future. I would have liked to have had Dr. Martin Luther King as a mentor. He was a great visionary. If he hadn't done what he did, we wouldn't even be having this conversation!
- **What would you have asked this person?** I would asked him, "What motivated you to fight for everybody?" He was not just fighting for African Americans. He was fighting for everyone.
- **What question would you like this person to ask you?** What can you do in your own context to continue this legacy? Wherever we are placed, we should be making a difference.

Afterthought

After the interview, we asked Estella if she felt she was a role model.

She shared that when her female students graduate, she would typically get a letter from them thanking her for being such a great role model and mentor. She then said she noticed they see in her a mentor and a role model, even if she did not develop any personal relationship with them.

There are people who just for being who they are, become exceptional role models. Estella is one of those.

Laureano Márquez

"*Humor is a mentor of justice and progress and of the wellbeing of people.*"

M. A. Capello and E. Sprunt, *Mentoring and Sponsoring*,
https://doi.org/10.1007/978-3-030-59433-6_14

A Glimpse

Sometimes, courage manifests itself in the shape of humor. In dictatorships and strong regimes and sometimes in democracies or pseudo-democracies, a humorist may risk not only his career, but his life when he mocks or alludes to mainstream or opposition politics, or ministers and presidents. This is the leitmotif of Laureano Márquez, a humorist, journalist, author and actor who has received many awards and has found a rich source of inspiration in Venezuela's political drama. He has exposed the flaws and ironies of political leaders in Venezuela. He cleverly finds the humor in the difficult, precarious and painful life that Venezuelans face every day in their impoverished country that ironically has a shortage of gasoline, despite sitting on top of the largest reserves of oil and gas on the planet.

He recognized his perspective was unusual during his university years. His classmates enjoyed his ability to imitate professors and peers. They secretly organized a surprise opportunity for him to do an improvised standup comedy sketch on his birthday on a real stage. That birthday party was the birth of the prolific and amazing career that balances Laureano's talents for writing and performing stand-up comedy.

Laureano excels in touching hearts and minds while provoking laughter from large audiences. He has performed live, televised and online in Venezuela and in México, Colombia, Spain, Bolivia and Argentina. His articles, which have been published in newspapers and online, prompted the late Venezuelan president Hugo Chavez to sue him.

A Personal Snapshot

Many of Laureano Márquez's articles focused on the decline of Venezuela's oil industry, which is a tragedy for a country whose budget depends on oil exports. Writing in a humoristic tone about dramatic events is not easy, but Laureano has mastered the skill, gaining the admiration of Venezuelans everywhere.

Laureano has been a voice that connects Venezuelans who due to the political disruption are now scattered around the globe. This interview was particularly poignant for Maria Angela. She had to reinvent her professional career in Kuwait and learn how to succeed in a very different cultural and work environment, because of the issues in the oil industry in Venezuela.

Talking to Laureano about his mentoring and sponsoring experiences was an eye-opener for Maria Angela. She learned about his approach to mentoring and sponsoring in a very different profession. The interview also revealed that Laureano is an extraordinarily humble human being.

He is the author of five humorous books and performs as the main attraction in thousands of theater and TV presentations. Laureano Márquez' work, like that of many humorists, is closely tied to his country. Now, with the worldwide Venezuelan diaspora, his humor has gone global, reaching Venezuelans wherever they are.

- A citizen of Spain and Venezuela, he was born in the countryside of Tenerife in the Canary Islands.
- He has 3.5 million followers on Twitter and 1.2 million followers on Instagram and is revered by Venezuelans everywhere.
- He was the 2010 Committee to Protect Journalists (CPJ) International Press Freedom Awardee.
- Author or co-author of 11 books and numerous scripts including: *"Historieta de Venezuela: De Macuro a Maduro," "Así es la Vía," "Código Bochinche," "SOS Venezuela," "No Pudimos," "Se Sufre Pero Se Goza," "Amorcito Corazón," "Las Actitudes Políticas: El cinismo y el Humor".*
- Accused by the late President Hugo Chávez, who successfully sued him for 3.8 million USD, of orchestrating a coup by writing a series of satirical articles, "*Dear Rosinés.*" The fine was paid with a national crowdfunding effort, in which thousands of people donated, to cover such a large amount.
- Script writer and comedian in "Radio Rochela," a pioneering and Guinness awarded comedy program, which was broadcast weekly without interruption from 1959 to 2005, more than 46 years.
- He has an extraordinary ability to play with the semantics and etymology of words.
- Holds a B.Sc. in Political Sciences from the Universidad Central de Venezuela.

The Interview

***Q.** Please tell us about your early years and your own mentoring experiences. Have you been mentored? How does mentoring work in stand-up comedy?*

First, I would like to thank both you and Eve for this conversation. For me, being part of your research is an honor. I consider myself to be a very fortunate regarding mentoring.

My formal education was not in acting. As usually happens with comedians who have a natural gift for humor, people around you start to appreciate that quality. When I was a student at the university, I was the person who would entertain the class by parodying our professors and classmates between classes and at lunch time. I considered doing these imitations the most natural thing in the world.

Celebrating one of my birthdays, my university friends arranged a small party at a place with a stage. Without advance notice, my friends asked me to stand up, go to the stage and do my imitations and jokes in front of complete strangers who were also in the venue! This launched my start in stand-up comedy.

One of the people who saw me at that venue was impressed and kept in contact with me. He worked for "Radio Rochela."[1]

One of the things I have learned, after so many years in this work, is that no matter what country or culture, everyone has a friend who "is very, very, funny," and would make a good stand-up comedian. TV and film producers know this very well, so having a friend advocate for you is not indicative that you will be able to break into the profession. This fellow was persistent, and he pushed the producers, writers and directors of "Radio Rochela" until Jorge Citino, the executive producer of the program, gave in and asked to meet "that genius comedian." That sponsor, whom I found by chance, was Manuel Gaitán.

During my interview, Citino asked what I did, and I told him I made imitations. He explained the cast was complete, and that they did not need any more imitators. I was already heading out, when Citino suddenly turned around and asked me, "*Hey, do you perhaps write scripts*?" I had never written a script in my life, but my answer was a bold and loud, "*Yes! Of course.*" That day I started working on Radio Rochela.

I was in a very difficult economic situation and needed income. I was terrified, but I started to write scripts. My most important mentor in this endeavor was Pedro Martinez, a professor at the Political Sciences School of the Universidad Central de Venezuela. He taught "Philosophical Principles of Politics" and was informed by my peers of my adventurous interview with the Radio Rochela executive producers. He asked me, "Do you know how to write scripts?" and said, "I have written quite a few, as that is my second job. Stay after class, and I will help you navigate this."

I could not believe how lucky I was. Professor Martinez taught me the basics, the strategies, basic schemes and the relevant best practices for shaping

[1] "Radio Rochela" was a pioneer Venezuelan late-night television sketch comedy and variety show, that ran from 1959 till 2010, which parodied contemporary Venezuelan culture and politics. It was mentioned in the Guinness Book of Records as the longest comedy television show in history, being aired uninterrupted for 5 decades.

scripts for TV. He was my very own Mr. Keating, of the "Dead Poets Society" film, urging me to "seize the day" or Carpe Diem.

I seized the day! My scripts were welcomed and used, and I was initiated into what became a life-changing experience. The team of comedians, producers, writers and support team members at "Radio Rochela" were truly integrated and all supported me in my initial years in this profession. I was invited to writers' brainstorming meetings, in which we ranked ideas for sketches, that then were distributed amongst us to be developed. People like Francisco Martinez, Miguel Angel Perez Belisario, Jorge Citino himself, and Carlos Sicilia, were not only writers of scripts, but mentors. I was the youngest and most inexperienced of them all. In particular, Sicilia was a very talented and ingenious script writer, who was always procuring ideas and recommendations for my work. Sicilia was a natural mentor for me.

Those brainstorming sessions were a school for me. The creativity and the strategies on how to use political or everyday issues going on in Venezuela to create comedy. Not just any comedy, but sketches that usually triggered self-reflections, and had a message. At the beginning, and for a long time, I wrote scripts for Radio Rochela without a salary.

Q. *When did you shift from writing to acting? Did you have any specific sponsors for your acting?*

Yes, and I would like to recognize in Ms. Maria Teresa Azuaje, a wonderful sponsor of my initial work. She was the right hand of Radio Rochela's Executive Producer Citino, and in turn would succeed him. Azuaje saw me doing imitations, and insisted Citino give me an opportunity in acting.

Citino and Maria Teresa noticed I resembled a presidential candidate of the time, Eduardo Fernandez. It was 1988, and there were two major political parties in Venezuela at the time, one, Acción Democrática was launching Carlos Andrés Pérez, with the theme "El Gocho pa' el 88" (a rhyme referred to his Andean region provenance and the election year), the other party was Copei, and their candidate was Eduardo Fernández, nicknamed "El Tigre" (the tiger).

"You look like Eduardo Fernández. Maybe you are our fellow for that imitation" they told me.

I was asked to prepare for a test with adequate makeup, hair and clothes for the impersonation. It was OK'ed, and I started impersonating Eduardo Fernández.

My counter figure in the sketches, impersonating Carlos Andrés Pérez (who won the 88 elections), was an experienced and revered actor and opera singer in Venezuela, Cayito Aponte. He would correct and help me saying,

"It is better if you look directly to the camera in this way," or *"Pause longer here after you say …,"* and other things like that. With the gestures, with the voice

I saw a lot of comradery, team spirit and support among all the actors Everybody would teach you, help you, correct you. Solidarity prevailed, and I think they all acted as mentors. The team at the time was filled with amazing talent: Cayito Aponte, César Granados Bólido, Nelson Paredes, Kiko Mendive, and Roberto Hernández!

I remember this collective support very vividly.

After a while, I became close to Emilio Lovera. Emilio became my mentor as well as my sponsor. He was contracted to organize the hiring of comedians for monologues at a seedy venue, the "Dance Palace." It was awful for other things, but it transformed itself at night, focusing on humor presentations attracting many people.

Emilio sponsored me to be a stand-up comedian in those humor nights at the "Dance Palace." He mentored and guided me in how to structure my presentations. These were my first truly solo stand-up comedy essays.

At this point, I had a stable job at Radio Rochela, as comedian and script writer, plus, I had my stand-up comedy presentations. Emilio Lovera showed his entrepreneurial vein and became my representative for contracts elsewhere.

Things evolved for the better and there was a much fancier venue for shows. It was "La Guacharaca," where top comedians had their shows. I was invited by Ben Ami Fihman who acted as another sponsor, supporting me as I was not yet, let us say it, … "famous!" This was an important milestone. He opened the opportunity doors for me. I then invited Manuel Gaetán to come to La Guacharaca. It was the first time I became a sponsor. In this case, we reversed our original roles. I paid back the extraordinary support I had received and was very happy. Gaetán coordinated the artists at La Guacharaca.

Q. *What do you like the most writing or acting your stand-ups?*

Writing is an intimate activity. Very personal and quiet. I like it a lot. It teaches you many things. I must research to write. And when the final book, or article comes out properly crafted, it gives me a great satisfaction. What I like about acting is the creativity. I really enjoy the monologues in comedy of humoristic nature.

After my time in La Guacharaca, I felt my career was on track. After that exposure, support and endorsement, sponsors kept coming. I have always been blessed with having open doors in front of me. When you acquire enough prestige in your profession, people ask for you.

I did everything in my power to be a serious person, but the only doors open to me were those related to humor including TV comedy. Those were not doors for which I searched. They were just wide open in front of me.

You may not know this, but I am naturally shy. The very idea of standing in front of an audience in public is for me a horrifying idea. When studying at university and in high school, I was afraid to give simple assignment presentations.

Q. *How is that even possible? I can't imagine a shy Laureano!*

Well, you are not the only one. I started elementary school in Spain at a school of the Marists brothers.[2] In recent years, someone from Venezuela told one of the old friars at my elementary school that Laureano Márquez was a very famous comedian. Br. Isidro simply answered, *"Well, you must be confused with another Márquez. My Márquez here, he cannot be that one. He was an extremely shy kid."*

I must comment that I didn't consider humor a serious activity. I ended up realizing that it could and necessarily had to be a serious thing. I could apply my political foundation in academia to humor. At work, and subsequently in my life as a comedian, whenever there was a need to prepare scripts about politics or political events, I was called in for the job. So, I became naturally focused, you may say specialized, in political humor. Some people still ask me if I use my university academic background in my work, and the answer is a big yes, as I work almost exclusively on political comedy, satire, and humor, grounding my work on political foundations.

Q. *You have 3.5 million followers on Twitter and 1.1 million in Instagram. Do you think you mentor others through your social media channels?*

People tell me, "Thank you for being our voice, and for saying what we cannot say." I believe humor is the voice of victims of power and power abuse. Humor is a mentor of justice and progress and of the wellbeing of people. I feel a huge responsibility when I write. As I have attested, what you write has consequences. That is the difference between writing and talking. What you write is, as Pontius Pilate stated, "*Quod scripsi, scripsi*" (the written, written is).

People attribute to me many things that I have not written. This past week many people started to congratulate me publicly for an article I did not write. I had to deny my authorship. I already incur many issues and problems with my own writings. Yes, when I see people trying to imitate what I

[2]The Marists Brothers are a catholic congregation dedicated the education of children and young people, with a preference for those who are the most neglected.

do, perhaps I act as a mentor without knowing it. Yes, people may perceive they are mentored by me through my social media.

Q. *When did you feel comfortable with your stand–up comedian label?*

Probably after "La Guacharaca" years. My goal was to engage in "serious" professions and "serious" activities. I realized that what I did was serious. It was my vocational calling, in which the Providence wanted to place me. So here I am, living my life with illusion, as Humor has chosen me, it told me "Come!" So, I obeyed.

Q. *After La Guacharaca period, did you have any other sponsors?*

Oh sure! Claudio Nazoa. Claudio and I asked the publishers at El Nacional[3] for space to publish humorous articles. They acceded to our request, but we did not have a permanent column. It was an occasional thing. Later, they launched a humor-dedicated section, directed by Pablo Brasesco, who invited me. So, Nazoa and Brasesco were my sponsors. In that role, as a contributor to the El Nacional humor section, twice I won the award for "Best Humorous Article of the Year".

These articles in El Nacional positioned me as a writer and enabled further opportunities that shaped my second career as a comedian. After El Nacional, I was hired by "El Mundo," another large-circulation national newspaper in Venezuela.

In 1999 Hugo Chavez won the presidential elections in Venezuela and in 2000 Teodoro Petkoff[4] founded the newspaper "Tal Cual" with an editorial stance opposed to the government of Chavez.

Teodoro Petkoff asked me to work on his new newspaper, Tal Cual, and I abandoned my paid job at El Mundo in order to work for an occasional salary at Tal Cual.

Q. *Why you did you do such a thing?*

Petkoff was an individual who had a giant stature in intellectual terms. You could not say no to Petkoff!

Teodoro Petkoff was a great mentor and sponsor for me. I started writing one humorous article per week. Then, he called me to tell me that he liked my articles so much, he preferred to use them as editorials. This was a huge thing! Teodoro was so important in terms of his thoughts and his ferocious

[3]El Nacional is a Venezuelan newspaper with national coverage.

[4]Teodoro Petkoff Malec (1932–2018) was a Venezuelan politician, guerrilla, economist and journalist. Minister of Planning, Petkoff launched the newspaper Tal Cual in 2000 and was its editor until his death in 2018.

opposition to the government. The newspaper was sold due to his editorials. After a short time, Teodoro asked me if I would always write the editorials of Tal Cual. That was a big step for me. Editorials in Tal Cual were posted in the front page. He had this big trust in me.

He was also an innovative pioneer, who realized a serious newspaper could have an editorial of humorous nature. Certainly, it was a first for Venezuela, and perhaps in the world.

My editorials brought many problems. The biggest was the famous "Querida Rosinés" (Dear Rosinés), an editorial I shaped in the style of a letter to the small daughter of Hugo Chávez. President Chávez sued the newspaper and me. We lost the trial, and the fine to be paid was about 4 million dollars.[5] There was a national crowd funding effort to collect funds for paying the fine. Leaders in all industrial and social sectors of the country, including ex-president Caldera, and other politicians, paid huge amounts to collaborate. It was heart-warming for me, to see this huge support from my countrymen and women.

Free Tickets

The Tal Cual editorial was paired with another phenomenon, "*La Reconstituyente,*" a humorous theater play. A small group of humorists, including me, decided to launch a play just for fun. We wanted that play to be related to the re-writing of the Constitution in Venezuela, a process that divided opinions in the country and was central to changing the political powers in place until then, but from the humorous perspective, since the whole process was almost surreal.

We chose a long weekend, Easter weekend, which in Venezuela is a 4-day holiday, to launch "*La Reconstituyente,*"[6] when the rental fees of the small theater we selected were the lowest. I must admit we did not have faith in our own project. The capacity of the venue was 300 seats. We decided to sell 100 tickets, and offer 200 tickets for free, as courtesy passes. We did not expect anyone to attend, and even estimated our individual shares of expected losses. But we went ahead with our project, for fun.

[5]The amount was 3.85 Million USD, at the official rate in Venezuela 2005, a non-heard amount for Venezuelan courts for a law suit referred to journalism.

[6]There is an interesting word-based humoristic game in this title. The Venezuelan Constitution was at the time in a process of being re-written, in what was a political limbo period, with a "Constituyente Parliament" (Constitutional Writing Parliament). Re-Constituyente instead is the name of a vitamin medicine.

- The first day, we had 10 paid attendees, and 4 courtesy passes.
- The second function, we had 20 paying people and 50 courtesy passes people.
- The third function arrived all holders of courtesy passes for days 1, 2, and 3, plus we sold all 300 day 3 tickets!

The theater was always full. It was an incredible success. That initiated a season for a play that survived two uninterrupted years of presentations. It was a play with high impact, and it gave me a high profile in political monologues.

Q. *Do you think your art and humor can be exported outside of Venezuela?*

Each country has its humorists, who interpret the country's political daily life in humorous codes that are very local. Spain has its humorists, who speak to the Spaniards. I have focused on Venezuela, because our country is experiencing terrible anguish.

Currently, I am living in Tenerife, Canary Islands, and have launched some acting and writing workshops here with the library. Also, "Casa de América," an institution that sponsors Spain-Latin America liaisons and activities, opened its doors to celebrate the launching of my recent book. In this case, Don Antonio Pérez Hernández, the Director of Casa de América has been a dedicated sponsor for me and my co-author.

It is not unthinkable or impossible that I could become involved in the Spanish humor, but currently, I am dedicating my time to writing. I am now enjoying a contemplative lifestyle, "*ora et labora*" (pray and work), as the priests say. Today, I have a conference call with some producers for a possible play in Perú.

Q. *You seem to like a lot of Latin phrases and historical similes. Why is that?*

I am not as literate as people think I am, but I am flexible and curious. I have an open mind. I need and enjoy studying the etymology of words, because I used that a lot for humorous purposes. Etymological analysis is a rich source for humor. I was lucky that I had excellent professors.

I think that I had more mentors than talent, and as more people trust me, that compels me to be more and more witty.

Q. *How would you mentor young comedians?*

I would focus my recommendations on how to make an audience laugh. I would talk to them about the nature of the relationship or perhaps the temporary bond that is established between a comedian on the stage with every person in their seat. How do you shape that magical effect, that one says something and makes people laugh? I would like them to ask me about the art of being funny and about how to craft a joke. About the limits of humor, and more importantly the purpose of humor.

Humor has purpose, spirit and a reason to exist, like a Polo Margariteño.[7]

Q. *Do you think mini-mentoring sessions can be effective?*

Yes, and perhaps that is what we need the most.

I am now shaping several workshops under the umbrella of a program I have called "La De-generación de Relevo" (Relay Degeneration, playing with the words). Some of them have become successful humorists. From this experience, I have learned the power of sponsoring someone. As some are successful, people think they are good because I mentored them, when instead I considered they are building their prestige with their own talent.

Many times, you can help a comedian with timing support. For a joke or a routine, you need an illuminating spark to ignite humor.

Q. *Do you think online platforms are good for mentoring?*

Yes, and in these days of COVID19, I have assembled sketches online with and for several humorists. But I much prefer in person contact.

Q. *Why aren't there many more women in humor? And have you had a woman as a mentor or sponsor?*

That is a very valid question. I think humor remains a macho terrain. I do not know if that is because exposing yourself on a stage live is dangerous because if your jokes are lousy, people will throw things at you. I am puzzled about this, because I think female response to stand-up comedy is faster and cleverer. Women have a faster sense of humor. Historically we have had exceptional female comedians including Lucille Ball and Carol Burnett who are masters of body language and ingenious comedy and recently, Ellen De Generes.

I do not think gender matters in crafting a humoristic career, but in Venezuela it has been very masculine. Also, in comedy writing. I remember only one female script writer, Yajaira González, from Radio Rochela.

Q. *Is there anything you wish you had done differently in your career?*

I consider myself to be very fortunate. I have huge gratitude for all I have received in life.

---

[7]Márquez is referring to a traditional folk music, Polo Margariteño. One of the most famous polos lyrics is "*El cantar tiene sentido, entendimiento y razón*", verses that every Venezuelan knows, and he used them to describe humor.

Q. Do you want to tell us an anecdote that resonated with you?

Cayito Aponte was a man with a prodigious memory. The first time I had to work with him was in my impersonation of Eduardo Fernández, in which he would impersonate Carlos Andrés Pérez. I was extremely nervous, as Aponte was already legendary, and I was just beginning.

They were applying the makeup on him, and he asked me to read the script for him. I read it once. Only once. And he memorized it immediately. He asked me to rehearse it and I got lost in my answers, even with the script in front of me! He instead, marveled me with all his phrases in place, already memorized and perfectly in the best comedy tone.

Another extraordinary thing that happened to me is that I was Master of Ceremony at the reception for the 80 years of a Venezuelan politician, Pompeyo Márquez. I was receiving the VIPs who were invited. Dr. Rafael Caldera, who was President of Venezuela at the time, was arriving. I saw it from afar. I had gained prestige imitating him, and I did not know what to expect. With tons of charism, he approached me, and the conversation went like this:

- (Laureano Márquez)—"*Welcome!*"
- (Dr. Caldera)—"*Are you Dr. Caldera?*"
- (Laureano Márquez)—"*No, Dr. Caldera is yourself*"
- (Dr. Caldera)—"*Ah OK. That's why you seemeed too tall to be Dr. Caldera*".

A Shared Selfie

- **Your favorite role model**: I have many, not only one. We already mentioned Lucille Ball and Carol Burnett. Then, there is that supreme genius, Charles Chaplin. In Venezuela, Pedro León Zapata and Aquiles Nazoa are emblematic, with a talent and compromise with humor. I admire Emilio Lovera, and his gift for humor.
- **One word to define your experience with mentoring**: Gratitude. First, because it is a word that expresses thanks, and because etymologically it comes from "*Gratia*", a Latin word that means a gift given by God. I think mentors are like Angels sent by God to support you in your path, so that your intelligence shines.
- **One word to define your experience with sponsoring:** Help.

- **Is there anyone that you wish would have been your mentor, but you never asked?** Yes. I wish I could have asked Jose Ignacio Cabrujas[8] for mentoring. I encountered him almost daily during my time at Radio Rochela in the hallways of the TV station, but I was too shy, and I never approached him.
- **Who would be an ideal mentor for you in this moment of your life? Any person, inclusive of a legendary, historical figure or other.** It may sound as a cliché, but I would choose Jesus Christ.
- **And what would you ask this Mentor?** How could I gain the grace of having the gift of wisdom, so that the right word would always be in my mouth.
- **What do you wish he would ask you?** If it was worthwhile. If the path was good.

Afterthought

We asked Laureano if he thought humor was related to a degree of intelligence. And he told us he believed it was instead a special kind of intelligence.

Of all interviews we conducted for this book, Laureano Márquez mentioned the largest number of mentors or sponsors—21, illustrating his humble character and grateful heart. Additionally, the collective spirit of mentoring and sponsoring of this sector.

He repeated several times, "Humor is a defect with which some people are born." We instead remain convinced it is not a defect, but a huge gift, and a special intelligence that shines in Laureano.

[8]Jose Ignacio Cabrujas (1937–1995), a Venezuelan playwright, theater director. He is considered one of the founders and innovators of the modern telenovela genre in Latin America and is called the "Maestro de las Telenovelas".

Shauna Noonan

"*I'm a better mentor, because I've had good mentors.*"

M. A. Capello and E. Sprunt, *Mentoring and Sponsoring*,
https://doi.org/10.1007/978-3-030-59433-6_15

A Glimpse

The US have seen a revolutionary change in their oil production schemes. Oil from shales obtained through fracturing has enabled the country to shift from being a net importer of oil to a net exporter. What were once considered unconventional resources are now included in the oil markets worldwide. Fracturing and artificial lift have had a huge role in this storyline. Shauna Noonan is one of the technical leaders who spearheaded not only the expansion and application of standards in this area, but also pioneered the inclusion of research and new technologies in the daily operations of top companies in oil and gas.

A technical leader, who is proud of her successful career in the technical and not managerial ranks, Shauna has led the application of new technologies and research pilot projects, particularly with regard to fracture productivity and conductivity, coupled geomechanics and fracture propagation tools, oriented perforating, flow control device characterization, and fiber optic completions. Her technical insights have developed to a point where she provides oversight and guidance for projects with global outreach and is charged with stretching production engineering to new limits.

- SPE 2020 President (serving on the Board of Directors 2019–2021)
- Hart Energy 25 Most Influential Women in Energy (2020)
- Director of Artificial Lift Engineering for Occidental Petroleum, Houston, USA
- Leading technical roles in Occidental Petroleum, ConocoPhillips, ChevronTexaco, and Chevron in Canada and USA, with global responsibilites as Chief of Production Engineering and Artifical Lift, Manager of Completion Technology, Wells Supervsior—Completions and Production Technology for Global Wells, and Staff Production Engineer for Completions and Production.
- SPE International Board of Directors (2012–2015)
- Chair SPE International Training, Programs and Meetings Board committee (2013–2015); SPE International Technical Director for Production and Operations (2012–2015)
- Chair of ISO 15551-1 Electric Submersible Pumps (2009–2015)
- ALRDC Artificial Lift Award for Exemplary Service and Contribution (2015)
- SPE Gulf Coast Section Regional Production and Operations Award (2012)

- SPE Outstanding Associate Editor: Production and Operations Journal (2010)
- Girl Scouts of America troop leader (2008–2015)
- McTeggart Irish Dancers of South Texas: President (2008–2010)
- Petroleum Engineer B.Sc from University of Alberta, Canada.

A Personal Snapshot

Shauna is the 2020 President of the Society of Petroleum Engineers, SPE, a not-for-profit organization with more than 160,000 members in 157 countries. This has made her a role model for young engineers worldwide. Many eyes are now seeing how much a technical person can achieve. We expect that new generations of production engineers, particularly women, will be inspired and motivated by Shauna, who is a champion of innovation. As the 2020 SPE President, she is sharing her ideals, goals and success story around the globe.

The Interview

Q. Please tell us about your own mentoring and sponsoring experiences.

I didn't start mentoring until I was a manager and learned how to do effective career coaching. Occidental Petroleum (OXY) required career mentoring of all high potential people, and as an engineering chief, I was a mentor for that structured program.

When I was selected SPE president, I started mentoring people outside my company, as people reached out to me via social media. Also, I was asked by people within OXY and SPE if I could mentor their daughters in engineering careers.

Q. How do you start your mentoring process?

It depends, but generally, I want preparation. I ask the mentees for their current CV, what their current role is, and what they want to do in the future.

I like to schedule an hour or an hour and a half for my first mentoring session. Then, it may be sessions of 30 min each.

Not all, but some of my mentoring processes are formal. One of my mentees had as one of her Key Performance Indicators (KPIs), to meet every quarter with me for a mentoring session. In some cases, I also do ad-hoc mentoring.

Usually, in first session I'm the one asking most of the questions and trying to find out what they expect to get out of the mentoring experience. I make a point of setting realistic expectations. I specifically tell the mentees that I am not there to tell them what they need to do.

I don't want to hear from the mentees in response to my questions, "*I don't know.*" If they have not done their homework and come prepared, no mentoring will help them. So, I send them back to do some deep thought.

Q. *Do you often mentor men or females?*

When I first started mentoring, I mentored men, because that's who I had in my organization. When I came over to OXY,[1] I realized they had a lot more females on the technical track, so I started mentoring women.

Q. *What kind of mentoring do you like to engage in?*

I tend to like to mentor people who want to stay on the technical track.

The ones reaching out to me over social media are always female professionals. I never turn one away. I also recognize those who are too shy to ask. So, I encourage them, by offering myself as a mentor.

Q. *Have you been mentored? Tell us about your own experiences as a mentee.*

I dedicated my February SPE president column to my mentors.

When I first started in Chevron, the company required you to be mentored. It was in the guidelines from Human Resources. During my early career years, Don Patterson was my mentor. My meetings with him struck fear into me. I learned to be very prepared, because he asked truly difficult questions. In retrospect, I admit he was providing good criticism and valuable feedback to me and other young engineers.

I also had a fantastic supervisor at ConocoPhillips, Michael Mooney. I always wanted to pursue a technical career track. He was instrumental helping me learn how to apply effective career coaching. He groomed me and he was great at mentoring me on how to be a good manager. He improved my managerial skills incrementally increasing my supervisory responsibilities and staff. I started with two engineers under my wings. Then he started adding more people for me to manage on my team. He was a fantastic manager.

Another good technical mentor for me was John Patterson. He was also my sponsor, and people considered me to be his protégé. While he was an

[1] Occidental Petroleum.

amazing mentor technically, he didn't understand work-life-balance issues, because his wife didn't work, so he was free to work late at night, and travel extensively. He could never understand why I didn't get a nanny to raise my children, to free my time to similarly work the same extended hours and take on unanticipated travel. I had personal obligations and a different schedule. For example, my daughters' had dance competitions that I wanted to attend to support them. All the family dynamics are different when you are a dedicated working mom.

Q. *Did you ever have a female mentor?*

No, I've never had a female mentor. All my mentors have been men. The most effective ones were those whose wives also worked.

I would like to highlight two women who served as good role models. One was Karen Draper, a woman who ran a very large pump company and was also a dedicated mother. She was very classy. At SPE, the electric submersible pump workshop is a major fund raiser for scholarships and she ran that for 20 years. She raised $200,000 for scholarships in a single event. She showed me how you can be a leader for SPE and for your own company at the same time, and she became my very good friend. When I had my first daughter, Karen came to the hospital to be with my husband and me, knowing that we had no family able to travel to Texas and attend the birth.

Carol Grande is my other female role model. She runs her own company. She became one of my very good friends as well.

Q. *You are SPE President. Did you have a mentor and sponsor for achieving this?*

Yes, a former president of SPE, Jeff Spath. He started out as a mentor and became my sponsor within SPE.

Q. *Do you like mentoring solutions online? Are they effective?*

If we can, I prefer mentoring in person, because what I really like about mentoring is to be able to focus on soft skills, which are now called strong skills.

Q. *Some people follow you on social media, do you consider what you post in social media is also a form of mentoring?*

It is role modeling, but not mentoring.

I prefer to mentor people in person, because I can read them better. I had a mentee from Trinidad and Tobago who was having a very difficult time at work and wanted mentoring via social media. We had several sessions of mentoring online. I posed critical questions to her. She decided to leave the country and the company for which she was working because the work environment was toxic. Now, I have been mentoring her for 18 months. She is thriving in Aberdeen and I am happy, because the mentoring worked out so well for her.

Q. *Now, please tell us about your sponsoring experience. Have you been sponsored to advance in your career? How? Have you sponsored others during your career?*

I would like to highlight that the first time I really thought about the need and the importance of sponsoring from my side was at a WIN event. WIN is Women in Energy, an SPE Committee whose goal is to facilitate the empowerment of women in Petroleum Engineering globally.

In that WIN event at the SPE Annual Technical Conference and Exhibition (ATCE) in Calgary, Dr. Christine Economides raised the issue of Sponsoring. I had not sponsored anybody and I didn't have a sponsor in my career. My only sponsor was at SPE, Dr Spath. This triggered in me a question: Why hadn't I ever sponsored anyone? I had a technical career, so maybe I was more focused on technical coaching, rather than career counseling or mentoring.

I always sought to help people rise on the technical track. Most or all the time, my mentees, even if I was encouraging them to stay on the technical track, switched to the managerial track. I must admit there are not as many opportunities in the technical careers for advancement, so maybe therefore I have not yet sponsored effectively anyone in the corporate realm. Also, as I moved from company to company, I did not build the necessary network for sponsoring people.

Q. *Wasn't the manager who tricked you into a supervisory role a sponsor for you?*

Well, I considered this was technical mentoring. I wanted to remain in my technical track and progress there. The managerial responsibilities given to me in completions were in another discipline outside my main one, which is artificial lift and production. I was not pleased with the promotion into an area that was not my main competency nor on the technical track on which I wanted to advance. In fact, this is one of the reasons I moved to OXY. I wanted to pursue my technical career.

Q. *If you want to stay narrow is it very hard to find sponsors?*

A big yes to that question. It is difficult. Mentors and sponsors become scarce as you move up the technical ladder. If you want to stay tightly focused on a technical discipline, it is difficult to find sponsors.

The person at the top of the technical ladder is a CTO—Chief Technical Officer—and a technical chief. There are very few people higher up than me. The CTO I'd like to emulate is Greg Leveille of ConocoPhillips, who came from the geosciences path. He is brilliant, but also strategically oriented. He stays engaged with people and showed me how to maintain my technical network. I learned from him to set goals at the beginning of the year for my inner and outer circles, to setup my "personal advisory board" and to prepare a communication plan with them for the year. Incredibly, he periodically made sure I was implementing and executing that year's plan and connecting with my contacts. It was an awesome way to mentor me for technical progression.

Q. If we want top managers to sponsor technical people, do we want to emphasize technical or soft skills?

You can be the best technical person in the room, but if people do not want to work with you because of poor behavior traits or you do not know how to communicate your ideas, you will not get to the top of your technical career track.

I almost put more weight on the soft skills.

You must know a little bit about everything now, when you are making real-time quick decisions. You need better soft skills, because you only have a short time to communicate your ideas to someone and must do it well the first time.

Especially for the young generations of engineers, I would put emphasis on their soft skills.

Q. *In your own experience, what are the questions that the mentees should ask you? Or in other words, what questions do you want to be asked in a mentoring session?*

I want them to ask, "What do I expect from this mentoring experience?" That helps to set realistic expectations. I tend to be the one who asks the questions in the earlier sessions.

Q. *Is it a heavy load on you to mentor people, guiding them about their careers?*

No. I don't feel the weight of mentoring on me, because the decisions are made by my mentees, not me. It is my mentee's decision. For example, I didn't tell my mentee from Trinidad and Tobago to leave. I asked her if she had looked around for opportunities and she moved to Aberdeen. If in Aberdeen the situation was worse, it was her choice, grounded on her own situation and she would then have to move on again.

A good mentor allows a person to seek out opportunities so that they can make a choice and lets them make the choice on their own.

I've had students reach out to me. People would ask what job should I do? Which job should I take? We'd discuss the pros and cons. Ultimately, they make their choice.

Q. *Do you become more proficient mentoring people with practice?*

I'm a better mentor, because I've had some good mentors.

Q. *Do you consider that single sessions, or mini-mentoring sessions are also effective?*

Over coffee is not good mentoring for me. I like a structured approach. I want to address critical issues, for example the mentee's goals for the year in relation to the company's goals. I want them to understand what value they bring to the company. Many times, they cannot articulate their value. If you don't know or your manager doesn't know, what kind of appraisal will you get about your performance?

I consider this kind of mentoring is the most useful, especially at the beginning of the work year. Key questions I ask are *"How are you going to demonstrate your value to your company? Do you know today what value you bring to the company?"* I help them structure their goals so that they are strongly enunciated and focused on value-adding. Then, I do periodic checks ins with them.

Another critical mentoring session is before the end of the year assessment, helping the mentees to realize the value they brought to the company. A recap. A realization of their accomplishments.

I focus on two elements: first, that they feel better about themselves, and second, that they can produce an effective elevator pitch. An elevator pitch is a 30 s speech of what you would tell a VIP of your company if you were in an elevator with them.

I have noticed that the mentees who feel they have been held back or passed over for promotions weren't setting good goals at the beginning of the year.

I turned one of my student talks for SPE into a presentation about what I tell someone when I start mentoring them. You must tailor it for different locations. It is easier to mentor someone after I've visited that part of the world and understand the culture (corporate and local).

Many young professionals who sign up for mentoring programs may not want to hear any kind of constructive criticism and just want a pat on the back.

As President of SPE, I found myself in many occasions asked about career path decisions. Those conversations might be considered mini-mentoring sessions.

Q. *Is gender or age important in mentoring?*

For technical mentoring, age is not an issue. For career path, it is. Information on older people's career paths is no longer relevant for many of the incoming generation unless you understand how the organization has changed.

Q. *More and more men have working daughters. Does that enable a better understanding and respect at work for women?*

The men who have been better mentors for me are those whose wives also worked. There is a study that surveyed 1000 managers in UK and US and looked at mentoring styles. The men who supported women better were those whose wives also worked. The field managers and lower level supervisors who supported me the most or showed me some level of respect were those who had daughters my age.

Every female needs to have a female on her personal advisory committee to gain perspective.

The generation coming up the corporate ranks has more women and men who had a mother who worked. However, numerous men currently in C-suite roles have wives who didn't work. This may be an issue with our oil and gas leadership, because their vision is different from that of the new generations. Fortunately, there are many men whose mothers have worked all their life. That is excellent role modelling which is transferable to the current corporate environment.

In my tours as SPE president, I always show a photo of my two daughters. My family is important for me.

Q. *Do you have a supportive spouse?*

I knew early on that I needed to find a partner who understood the oil and gas side of things and what it was like when I was at a rig. I wanted to choose a partner so that our careers would be on par or mine a priority, where I would be the major breadwinner. I ended up marrying my slick line operator. I didn't marry until I was 28. It was considered kind of late by my parents. My mom hinted it was time to get married by buying me a china service!

When I was transferred to the US, it was only supposed to be a two-year hitch. We quickly realized how tough it was to live off one salary and he couldn't get a work visa, so he went back to school and got a degree in electronic engineering. Early on, we both traveled a lot and would be flying back home frequently. For years, our daughters only saw one parent at home. My husband is now a vice-president in his company. He has been extremely supportive all the way through.

Q. What do you wish you would have done differently in reference to mentoring or sponsoring, as a mentor or as a mentee?

I wish I had done more mentoring earlier on. I'm realizing now how rewarding it is to be a mentor and wish I had influenced more people. I regret losing some of the people, I tried to mentor to the managerial track. I wish I had done a better job of demonstrating how to succeed in the technical track.

Oxy was not creating many subject matter experts from within. Now, we have created a completely technical track organization. I think it is working. We just launched it about two and half years ago and we are getting people up through those pipelines.

Q. Did you enjoy any mentoring/sponsoring experience?

I had two wonderful professors who really engaged female students, Dr. Farooq Ali, and Dr. Anil Ambastha. Dr. Farooq aimed to mentor and sponsor female students. He really wanted his female undergraduates to do their MS under him. Helen Chang was one of his MS students. He was disappointed that I did not pursue a MS.

A Shared Selfie

- **Your favorite role model**: Karen Draper
- **One word to define your experience with mentoring**: Expanding.
- **One word to define your experience with sponsoring**: Limited. Because I have had a limited number of people who have sponsored me, and I have been very limited in helping others.
- **Who you wanted to be your mentor/sponsor, but you never had the chance to ask?** My CEO, Vicki Holub.
- **Who would be an ideal mentor for you in this moment of your life?** Helge Haldorsen with Equinor. He was one of the few SPE presidents, who continued to work in his technical role as SPE President and continued to do that when he rolled off the board.
- **What would you ask this person?** I would ask him, how I can use my experiences as SPE President in any future roles I have within my company?

- **What would you like this mentor to ask you?** How do you think you can use your experiences as SPE president in future roles that you have within your company?
- **What is your elevator speech?** Every rung on my career ladder all the way up to being engineering chief and SPE president had to do with my involvement with SPE. I moved up because I was writing technical papers for SPE. I wrote not because I wanted necessarily to get them published, but because I wanted my upper management to realize the value I was bringing to the company.

Afterthought

"*Having an SPE membership is like having a gym membership. It doesn't mean you will get fit. You have to use it to gain the benefit from it. I gain benefit from SPE, as I put my membership to work*". Shauna uses this phrase frequently. It conveys a powerful message.

We interviewed Shauna during the corona virus pandemic, while she was working from her home in Houston, USA. Her year as SPE President has certainly been during a difficult time with rescheduling of many technical events and activities all over the globe. Both the SPE staff and members are learning to work and volunteer in new ways. Shauna's tenure as President will probably be remembered as the year in which a SPE woman president changed workflows to focus on getting the job done remotely, leading global online collaboration of volunteers.

Dr. Elijah Ayolabi

"*I have a desire to assist younger people to become leaders in society.*"

M. A. Capello and E. Sprunt, *Mentoring and Sponsoring*,
https://doi.org/10.1007/978-3-030-59433-6_16

A Glimpse

Not all mentoring or sponsoring processes are the same. Cultural context and the personal styles of the individuals involved make a difference. Nigeria is a country with a large population, and that on the surface appears to be highly networked, communicative and open. You might assume there is a good framework for implementing mentoring and sponsoring. However, in Nigeria, hierarchies are important, academics are highly revered, and deference to experience and rank in the corporate ladder produces a deferential approach to mentors and sponsors.

We had an opportunity to discuss the mentoring and sponsoring in the Nigerian context when we interviewed Dr. Elijah Ayolabi, Vice Chancellor of the Top Mountain University, in Lagos. Dr. Ayolabi is an accomplished academic leader, who has contributed greatly to the advancement of geosciences and, in particular, to applied geophysics at several universities of his country. Dr. Ayolabi, the author of four technical books, has mentored thus far more than 112 students through the process of writing their doctorate and master level theses. His life has been dedicated to forging the professionals of the future and hopefully many more.

Dr. Ayolabi he earned his Ph.D. in Geophysics in 1999. His experience as a lecturer and professor is impressive. He has held positions in a multiple research and teaching organizations, including the University of Lagos, the Shell Center of Excellence in Geosciences and Petroleum of the University of Benin, the Federal School of Dental Technology and Therapy (Enugu), OSUSTECH—Ondo State University for Science and Technology (Okitipupa), the Osun State College of Education (Ilesa), the Ogun State University (Ago-Iwoye), the University of Agriculture (Abeokuta), and Mountain Top University (Lagos).

He has taught thousands of students thus far, often in large classes. In geosciences in Nigeria, especially in geophysics, classes often have more than sixty students. This has been true for the more than 20 years he has been teaching.

Especially notable aspects of his cv include:

- Master and Ph.D. in Geophysics, University of Ibadan (Nigeria), in 1994 and 1999 respectively.
- Ogun State University grant for Ph.D. Research (1998/99).

- NAPE/SHELL Awards for Effective Development of Quality Education in Nigerian Tertiary Institutions (2003), NAPE—AAPG Young Professional Faculty Adviser of the Year 2013, AGSC Special recognition award (Conference Outstanding Contribution) 2013
- SEG—Society of Exploration Geophysicists 2015 Honorary Lecturer (HL) for Middle East and Africa.
- University of Ibadan 2018 Lifetime Achievements Award (2018)
- Fellow, Institute of Management Consultants (2019)
- Vice-Chancellor, Mountain Top University, Lagos, Nigeria.

A Personal Snapshot

Maria Angela had the pleasure of meeting Dr. Ayolabi when she visited Nigeria as an SEG Honorary Lecturer. She was already familiar with Dr. Ayolabi's reputation as an academic because of his participation in multiple SEG (Society of Exploration Geophysicists) committees and from his NAPE (Nigerian Petroleum Annual Exhibition) papers. She found him to be not only an outstanding professional but also a very gracious host: When he organized the schedule for Maria Angela as lecturer at the Table Top University, he included a tour of the Music School Department, with an impromptu recital and meeting with the master degree music students, who compelled Maria Angela to play the piano for them. As unexpected as this was, it dramatically illustrated the style and improvisation of a culture Maria Angela discovered was spontaneous, caring and open to foreign perspectives.

Maria Angela surrendered to the warm welcome and after playing a few pieces at the piano, a formal group photo was taken. After this unique welcome, she continued her personal discovery journey, that included facing for the first time an audience of 300 attendees in Africa, and a gala lunch with the top officers of that university. All organized with a attention to detail by Dr. Ayolabi.

The Interview

Our interview session via teleconference encountered technical glitches that were surmounted because of Dr. Ayolabi's enthusiasm for sharing his learnings about the topic. He is passionate about mentoring, but avoids sponsoring, because he is very conscious of the importance of transparency and need for fairness in all processes in Nigeria. As a high-ranking person in the educational system of his country he must role model transparency and fairness. Fair processes preclude Nigerian academics from sponsoring individuals based on their personal impressions and convictions. To promote transparency Nigerian academics, including Dr. Ayolabi, must strongly support ranking and selection processes and work to ensure that they are fair and free from any personal influence. It is of paramount importance that the processes can be trusted.

After several futile efforts from our side to explain that sponsoring is frequently present in human interactions, we learned, that in some cultural contexts, it is more important to ensure fairness is in place because the whole society is moving away from sponsorship after suffering from an excess of sponsoring, which has tarnished the country's reputation for fairness in appointments and selection processes of leaders in public and private organizations.

Our discussions with Dr. Avolabi enabled us to understand that mentoring is needed everywhere, but sponsoring processes must be carefully scrutinized to ensure they are equitable, fair and harmonious with the local cultural and ethical values.

Q. Please tell us about your own mentoring experiences.

Mentoring involves providing guidance or direction for someone in life and may involve showcasing your own life and successes, in order to provide models for others to emulate. Generally, for me, mentoring starts with a one-on-one interaction with the mentee, like a discussion. In general, the mentee would have to look for the mentor, and where necessary, the mentor should look for the mentee. The one on one interaction with the mentee will enable the development of a sort of chart for a career part, perhaps also, provide a vision about a life style, aspiration or vision. Occasionally, it would serve to lead the career.

Q. Have you mentored many people? Do young professionals reach out to you to receive mentoring? How does your mentoring processes start?

Yes, absolutely! Many young professionals and students do reach out to me to be mentored. I have supported several mentees and brought them up to the lime light. I think I have mentored over 100 people, of all kinds: Students, faculty members and outside the academia, also young professionals.

Q. *Now, tell us about sponsoring. Dr. Ayolabi, have you sponsored others during your career?*

Yes, I had several opportunities to sponsor people one way or the other. For example, as a Professor, references are very important for students and for peer professors. I have given references for several people, that I considered were outstanding. I also propelled key individuals who were keen in pursuing their careers outside of Nigeria, by serving as an endorser of their academic work, with letters of recommendation. From the feedback I constantly receive from them, this action was a key element for their careers, and I know realize this is a type of sponsoring that I was able to activate.

Q. *Have you been sponsored to advance in your career? How?*

Fortunately, I have also been sponsored in the course of advancing my career, in the sense that I have been supported in my career. My studies were supported.

Q. *What were the initial sponsoring loops you were involved with?*

I would say that I have always been sponsored by many supporters of my work since I was a student. When I started my MSC and Ph.D. programme, my supervisor was able to connect me with another Professor who provided me the research equipment. Without this, it would have been impossible for me to even attempt a graduate program! I was not only fortunate but blessed with the opportunity of reaching out to this opportunity.

Similarly, the grant support (funding) for my Masters and Ph.D. programs was provided through the help of my supervisor. This economic support for me was a kind of sponsoring that was foundational to advance my career and my life through the years, with an academic degree that would open further opportunities.

As a way to give back to my community, I provided support to others, for instance, by serving as a reference for those seeking a scholarship and for those seeking admission to programs in other countries. I have also supported junior faculty including some at University of Lagos and Mountain Top University, helping them to accelerate their career progress. I have also sponsored several for job placement. Nigeria is a very competitive environment for our national young talent, and it is difficult to pursue advanced studies or gain a job employment offer, and I was pleased to support key outstanding individuals deserving sponsoring with my support. This is the sponsoring I prefer: the one grounded on quality of the individual, that propel my wish to endorse them.

Q. *"You are vice-chancellor. Who supported you to reach to this high rank in the academic world of Nigeria?"*

This question, as originally formulated, triggered a long answer from Dr. Ayolabi, explaining to us the need for transparent processes in Nigeria, and the importance for all involved to make sure selection and promotion processes are systematic and free of from influences of any kind. At our end, we explained that even if fair, the selection processes, in every culture generally in some way involve endorsements and someone to step in with a good word about the best candidates, even if it is in internal meetings that are secret to the external world. We rephrased our question, to match the cultural context, and the obvious pride of Dr. Ayolabi to have been selected in transparent processes.

Q. *Let us rephrase the question to: You are vice-chancellor. Who do you think may have supported you to reach to this high rank in the academic world of Nigeria?*

It is a competitive bid, and it is very difficult to know who spoke well for you. However, I imagine the proprietor of the university must have provided support for the selection, as he was the head of the selection committee.

Q. *Let us shift gears, and focus now on your own experiences as mentee or sponsored person. Do you remember any mentors or ponsor?*

Yes, vividly and I would like to mention Prof. Tolu Odugbemi, the former Vice-Chancellor of University of Lagos. He was the Vice-Chancellor while I was the head of the Physics department. His administrative style, commitment to duty and trust in divine ability were great incentives to me as the department head under him. We received generous support which led to the birth of Geosciences department at the University of Lagos.

It was an unforgettable experience, which led me to admire him in a great deal. I consider he put me under his wing, and I grew a lot because of this, professionally and personally.

Q. *Many times, the advantages of a mentoring liaison lie not in the person who provides the mentoring, but in the one seeking that liaison. In your own experience, what are the questions that the mentees should ask you as a mentor? Or in other words, what questions do you want to be asked in a mentoring session?*

Mentoring is like training for young professionals. I do not give a straight answer at the beginning. With students, we start with practical field experience, history and case studies from which they learn the steps that should be followed. Usually, I expect the mentee to ask me about steps to become great in life or to ask me how I got to where I am today.

Q. *Would you share with us any difficult mentoring experience?*

Oh yes! Often, I consider the greatest difficulty is to find the available time. And not only the mentor's time, but also the availability of the mentee. How hard it is depends on the mentee. There are very easy ones, but some are difficult.

Q. *Why they are difficult?*

Some are difficult in the sense that most of the time they are not available and when they become available…they are not ready to go through a long time of mentoring, as they have not analyzed where they want to go, or do not know what questions to ask. You must come prepared to a mentoring session.

Q. *Dr. Ayolabi, your observations propel me to ask you if you think single sessions, or mini-mentoring sessions can be effective?*

In my opinion, mentoring requires more than one session. I think that single sessions cannot be effective. Mentoring must be done for an extended period.

The mentor and mentee need to meet from time to time, depending on the availability of the two. The duration is not a fixed period, but not a single session. The period of mentoring may be 6 months, and often 1 year to 3 years. For those in a Ph.D. program, mentoring by a good mentor, one committed to every mentee's success, generally takes 3 to 4 years.

(Here, we noted that Dr. Ayolabi seems to be indicating that the students are mentored throughout their time as a student with Ph.D. students being mentored for a longer period, because it takes them longer to get their degree.)

Q. *Do you think mentoring has to be in person, or are on-line, digital mentoring solutions equally effective?*

Both are acceptable. Digital should be used to complement the one-on-one interaction.

Q. *Some people had a special influence in our careers. Please tell us about your own experience as mentee. Who mentored you in a memorable way? Who sponsored you?*

I had three mentors who impacted my life positively. When I was much, much younger, my maternal grandmother mentored me, influencing me to become hardworking and committed to my work. She was definitively a huge influencer for me. Her lifestyle, hard work, and commitment impacted my real life. When I was a younger person, my maternal grandmother taught me how to be a person and influenced my attitude about work.

The second person I would like to mention was my MSc and Ph.D. supervisor, who was from India. Dr. S. C. Garde. His attitude to supervision influenced what I do until today. My advisor allowed me to just go and sort things out, in a practical manner. My advisor's attitude influenced my commitment, ambition and what I do now. He took mentoring as a valuable interaction. He influenced my commitment and attitude. Normally, from time to time, he would provide me with his feedback. We shared experiences, progress and challenges together and the feedback proved insightful to trigger step changes in me.

I was also influenced by the proprietor of my current University—Mountain Top University—Dr. D. K. Olukoya. I adopted his way of life and attitude about mentoring younger people. This mentor has commitment and passion for developing younger people in a measure that is bigger than anyone else's I know. His desire is to see young people achieve positions of importance and relevance in the society. This instilled in me a desire to similarly assist younger people to become leaders in society.

Q. *Is gender or age important in mentoring? And in sponsoring? Females mentoring females, or more-experienced professionals mentoring young?*

Ayolabi thought for a while, and then, he answered, Generally, to me, gender is less important. However, mentoring is more for professionals than younger students. About age, let me emphasize that generally, age is not a barrier, but may be a critical factor, especially if the mentee is much older than the mentor. Those interactions may not be good. I have mentored female students with excellent disposition and integrity and today they are important pillars in our society.

Q. *What do you wish you would have done differently in reference to mentoring or sponsoring, as a mentor or as a mentee? as a sponsor or a sponsored professional?*

What I would have loved to do is to be more decisive and committed. Not differently, but more decisive in some of the things. I would have added more force to support my mentees.

Q. *Did you enjoy any particular mentoring/sponsoring experience? Why? Tell us more about those joyful occasions!*

Yes, mentoring by my Ph.D. supervisor was a memorable one that cannot be forgotten. When I started my research work, he took me to the library, and we started searching for references related to my research together. After finding two or three publications he told me to continue to search until I was able to find what I was looking for, and he left me there.

He called it serendipity.

That was how I was initiated into the world of research. I keep working in research every day.

As a mentor, I have similarly taken mentees to the library or provided some initial references and told them that until they find the solution, they cannot get out. I do this to allow them to do it themselves and understand the need to commit to finalizing a task, to research.

In the field, frequently, the mentee will not enjoy having to figure out for themselves the solutions, or the outcome of a research, but this approach allows students to make mistakes and correct them, so that they don't repeat the same mistakes again.

I have also done this with my mentees, taking them to the library in the same way as my supervisor did. Beyond that, what I do most of the time is to take them to the field, explain what is to be done and allow them to do it themselves. Then, we go back and fix any mistakes. Clearly, the mentee will not enjoy it, because is not easy to acquire data. I try to do this for electrical resistivity methods. It is important to make sure you do not let allow your bare hand to touch the naked wire on the electrode while transmitting current or you will experience an electric shock. I have let them touch the naked wire on the electrodes while transmitting current. After they feel the electricity, they are more careful not to make the same mistake again and they have a better understanding of the importance of safety in the field.

A Shared Selfie

- **Your favorite role model**: Dr. D. K. Olukoya, my proprietor, the owner of the private university where I currently work would fall into that category.
- **One word to define your experience with mentoring**: Interesting.
- **One word to define your experience with sponsoring:** Impactful.
- **Who you wanted to be your mentor/sponsor, but you never had the chance to ask?** Time is gone for this. Not anymore.
- **Who would be an ideal mentor for you in this moment of your life?** As a Christian that would be Jesus Christ.

- **What question would you ask this mentor?** Why did humanity conspire against Him, knowing that He was trying to help them?
- **What question would you like this mentor to ask you?** How has life been on earth particularly in Nigeria? How is life in Nigeria compared to developed nations?

Afterthought

Given the importance of understanding Dr. Ayolabi's valuable insights on the hazards of sponsorship and its potential negative impacts on society, our interview with Dr. Ayolabi was supplemented with a rich exchange by email to clarify critical points.

We were glad we include Dr. Ayolabi's insights in our compilation, as those greatly help in understanding some of the many nuances of multiculturalism in mentoring and sponsoring.

Johnny Di Francesco

"Sponsoring is to support people to reach their goals."

M. A. Capello and E. Sprunt, *Mentoring and Sponsoring*,
https://doi.org/10.1007/978-3-030-59433-6_17

A Glimpse

Australia is a very special continent, where cultures have merged in interesting ways, creating a platform for success for those who seize an opportunity. One of the most successful Australians is Master Chef Johnny Di Francesco, owner and founder of 400 Gradi, an iconic Melbourne institution that offers dishes rooted in the authentic Neapolitan tradition. 400 Gradi is now a global enterprise. In Australia it has locations in in Brunswick, Melbourne (including at Marvel Stadium, which is Melbourne's major sporting and entertainment hub, and at Rod Laver Arena, home of the Australian Open), Essendon, Eastland, Yarra Valley and Corwn Casino. Di Francesco also has venues in New Zealand, Kuwait, Bahrain, on board cruise ships, and has plans to open others in the USA, and Europe.

Di Francesco's margherita pizza was recognized in 2014 as the best in the world at the Parma World Pizza Championships in Italy. Winning a best pizza in the heart and cradle of pizza is nothing less than outstanding. Then, in 2017, his pizza was recognized as the best in Australia. In the 2018 and 2019 editions of 50 Top Pizzas, his restaurants were declared the best in Oceania.

- Australian of Italian origin.
- During his first year at university while studying engineering, he found himself daydreaming about the physics of pizza dough and realized his career should be in the culinary sector.
- He was crowned as the World Pizza Champion in Naples, Italy as the best of 600 competitors from 35 countries.
- "Best in the World" award at the Parma World Pizza Championships for his margherita pizza in 2014 in Italy.
- Founder and owner of "400 Gradi" a franchise chain of restaurants born in Australia, that offers Italian food in multiple countries.
- "Best Restaurant in the country," Australia, 2017; "Best Pizza Restaurant in Oceania," 2018 and 2019.
- A regular invitee on Australian television, he won the 2017 season of MasterChef Australia, and participated in "The Mentor" in 2018.
- Owner and Executive Chef of more than 10 restaurants, and now launching "400 Gradi Pronto" and "Zero Gradi Gelateria," which are fast food and ice cream franchises.
- Was the Australasian Principal in 2012 and 2018 of V.P.N (Vera Pizza Napolitana), an international association dedicated to protecting and promoting authentic Neapolitan Pizza.

A Personal Snapshot

Expanding his business beyond Australia required a vision of the future and courage. We wanted to understand the roles of mentorship and sponsorship played in his success.

Including Chef Di Francesco in our compilation adds insights from a very different business sector. We found perspectives fascinating and believe that they provide valuable lessons in other career paths.

The Interview

***Q.** Please tell us about your own mentoring and sponsoring experiences.*

Until I was 23 years old, I was mentored by my father. I initiated my entrepreneurship very early in life and employed my father. He was very good at his craft, and he mentored me, shaping the way I look at business today, with a focus on hard work.

"*You need to work in your business, not for your business,*" my father said. He meant that I had to know everything related to my business. I had to be engaged in all the details to be able to master and coordinate everything from the top down.

How can you drive your business if you do not know every single element? He taught me how to work with this sense of ownership, and I adapted his principle to all that I have done. Even now that the business has grown very large, I still rely on the same principles, hard work and perspective.

It is like driving. You do not understand how to drive, but you do it. Most chefs are great at cooking, but not at running the operations of a restaurant. Just because they study to be a food technician or a good cook, they will not automatically become a good manager. It is not like that. To become a good operator is something else. It is running the whole business.

***Q.** How did you become a manager?*

I was very lucky when I met Luigi. He was a father figure who became my best mentor. He had a wonderful perspective, and a vision of the whole operation of the restaurant.

He would always say, "You need to understand every part, every aspect of your business. If you face problems, then you will know how to fix the problems."

You then have the solution right away for any issue, 90% of the time. When people confront me with problems on the administrative side or procurement or the kitchen or waiter recruitment, I can answer, "*This is what we need to do.*"

Becoming a good operator depends on how many mistakes you have made.

People tell you, "*You are very successful*" not realizing that it is because I have made more mistake than others have.

I adapt and with that adaptation, I understand how to deal with difficult situations.

***Q.** Have you mentored many people?*

Countless! I have mentored staff, cooks, and admin people. I like to mentor, and I have mentored my children.

***Q.** Your restaurants are spread globally. How do you manage to mentor and direct your operations across cultures?*

To mentor people from different cultures, you need to understand their level of understanding.

Being a business owner is like being a psychologist without the degree. You need to know how to help, direct and assist at different levels. Business people, chefs, helpers, investors, waiters, and so on, all have different knowledge. You must understand their portion of the story.

***Q.** So, do you consider yourself a translator?*

Absolutely, I move across cultures and disciplines, talking to each one within their own perspective. That is to translate.

***Q.** Now, please tell us about your experience with sponsoring.*

On a personal level, here in the Middle East, I had a friend, who worked for the Al-Shaya Group. He used to visit Australia and was scouting for new businesses and new concepts to bring to the Middle East. He always said that "*your concept is good.*" He believed in my brand. He said that people in the Middle East would be attracted to my idea. He opened doors for me to the Middle East. He put me in contact with a group of investors and businesspeople who started the franchising of 400 Gradi, so that today, these restaurants that you see in Kuwait, are a successful reality.

Sponsoring happens in the business all the time. When we have a position, instead of going out for recruitment, we promote and test internal talent. I ask, "This person has exceeded expectations in her tasks, why don't we give her a try?".

My managers would say "No, it will not work, too much to learn," but I advocate for people, especially within the business. You see people who have the drive and the passion for the brand. This is pivotal, and it would silly not to provide them with an opportunity.

I will give you one example. One of our Executive Chefs in the Group literally started washing dishes.

The sponsored people will sponsor others. In addition, they will understand the character for which I am looking and the dream. Their achievements will be larger if they sponsor others, who in turn follow the same dream. When they sponsor someone, they will look for that kind of profile.

If you give someone an opportunity, he will want to replicate it.

There is no excuse if he or she is now an Executive Chef. An executive role does not imply the person has changed. He or she wants people around like him or her.

In my opinion, when you bring in someone from outside the group, generally they try to change the culture. It is disruptive. Our culture is so strong, they cannot get in.

Q. *What do you think about leading by example? Is that mentoring?*

I convey passion for what I do. I guess you may call that "leading by example."

One guiding principle of mine is to not ask someone to do something I would not be willing to do myself, even if that is cleaning the bathrooms.

Q. *What are the questions that the mentees should ask you?*

One question would be, "Why did I get into business?" I think this is very important. Many people believe they know what they need to do in a business, but they do not understand the "why."

There are 90% of people who will say to you, "Because I am passionate" or maybe, "Because I want to make money." Only 10% will know the "why." Your own "why."

Many will have the "what" and "how", but they do not understand the "why." People mix these three different aspects of business. If I go back now to when I was 22 or 23 years of age, my "why" was completely different than my "why" today.

To further explain, most people would be thinking, "*I will make more money.*" Yes, that is what is going to happen, but not the "why."

Another question would be, "*What is your 5-year or 10-year plan?*" Most people don't know what they want to achieve in the next years. That question is important.

Also, it is important for a mentee to ask, "*How do you know your end-goal?*"

I work in reverse, starting with my end goal. I define the steps needed backwards to achieve what I want.

If you are analyzing your goals for what will happen in 12 months or 2 years, the steps change. I will share one example. If you want to buy a house next year or in 5 years, your steps change, and you need a plan for that. So, it is key to set the end goals and the timeframe.

Every single day, I know what I must do. It is difficult. I know I need to save money. But how much money? It comes from planning backwards.

Q. *Do you think single sessions, or mini-mentoring sessions are also effective?*

I mentor daily. Mentoring is part of my daily routine. Many times, I receive mentoring, other times, I give mentoring. This is important. Sometimes, you are not aware this happens. My people sometimes do not see these as mini-mentoring sessions, but they are, and I believe those are effective.

They say something, and the trick is to listen, to pay attention.

The trick is to be present. Many times, people just hear, but they are not in it, not present. Listening is a skill.

Q. *Do you think mentoring has to be in person or are on-line, digital mentoring solutions equally effective?*

I think mentoring can occur on any platform, phone conversations, email, in person online.

The mentees must be invested in learning and improving what they do.

Q. *What do you like the most? Is it supervising the chefs or the business loops?*

Everything is very important.

Q. *Are gender or age important in mentoring and in sponsoring? Females mentoring females, or more-experienced professionals mentoring young?*

I went into business very young. I will never forget, this a true story. I was sitting in the front of my first business. I was only 19 years old. I was waiting for an electrician.

An elderly lady walked down the street. She stopped and asked me, "Young man what are you going to open here?" and then "How old are you?" I answered 19.

She said, "*Good luck, and I hope you understand that for the first 10 years you will not make any money.*"

I thought, "*What is she talking about?*"

But I can assure you in the first 10 years, I made zero money! I had to finance credit card debts with other credit cards. Not everyone knows how difficult and complicated it was for me to achieve success.

Q. *Did your resilience play a role?*

People talk about it, but few experience resilience. I did not come from a wealthy family. They were struggling. From a young age, I would sit in my bed without dinner and think, "*I am never going to live like this again.*" I saw my friends in a better situation.

My kids appreciate what we have done. My wife and I have not lost track of our humble origins and raised our children with the understanding of how fortunate they are. This past Christmas, they did not ask for any presents. They know there is no need for extra things in life.

I talk to my three children about the hardships in life. I think it is useful for shaping their own perspectives.

Q. *What do you wish you would have done differently in reference to mentoring or sponsoring, as a mentor or as a mentee?*

I wish I was more present. Luigi was indeed like a real father.

I was a judge in a pizza competition, and Luigi called to tell me he was not feeling well. It was too late. I could not do anything else for him. He died, and I still miss him very much. He was my best mentor.

A Shared Selfie

- **Your favorite role model**: I don't have one.
- **One word to define your experience with mentoring**: Leadership.
- **One word to define your experience with sponsoring**: Opportunity.
- **Who would you have liked to ask to sponsor you, but never asked?** Richard Branson.[1]
- **Who would be an ideal mentor for you in this moment of your life?** Albert Einstein. What drove him to think about things in an irregular way. He had a tremendous intelligence. What did he do in his daily life that

[1]Sir Richard Charles Nicholas Branson (born 18 July 1950) is a British business magnate, investor, author and philanthropist. Founder of the Virgin Group, overlooking more than 400 companies. Branson has an estimated fortune of 5.1 billion dollars and initiated his entrepreneurship at 16 years of age.

triggered his approach to thought? Many people are smart, but of those, only a few selected ones have the natural ability to use their mind in a capacity not understood by the rest of humanity.

Afterthought

All segments of activity require mentorship and sponsoring. We realized that what we do in the energy sector and in academia, mentoring people constantly also occurs in other vocations including the food sector.

The opportunity to interview a chef of the caliber of Di Francesco provided better understanding of the multitasking and pragmatism of an entrepreneur, who does not hesitate to provide opportunities to those who share his spirit, because that is more important than having specific skills or competencies. The keys to success are the attitude and style of doing things, which for Di Francesco means having his teams provide the best food and service.

C. Susan Howes

"Sponsoring is in fact advocacy."

M. A. Capello and E. Sprunt, *Mentoring and Sponsoring*,
https://doi.org/10.1007/978-3-030-59433-6_18

A Glimpse

Susan is an iconic role model. She has been a tireless volunteer for numerous organizations and a leader in multiple companies. Her accomplishments at Anadarko, Chevron and Subsurface Consultants Associates (SCA) demonstrate the power of her tenacity, innovation and networking in all aspects of life and business. She has worked in reservoir management, recruitment, training and on-boarding of new hires.

Her volunteer work has had global impact of great significance for future generations. With her seemingly endless energy, Susan has spent countless hours leading and supporting high-impact initiatives in not-for-profit organizations including the Society of Petroleum Engineers (SPE), the Girl Scouts, the University of Texas (UT), the Colorado School of Mines and the Society of Women Engineers (SWE). She has been the igniting spark for numerous initiatives that support individuals in Petroleum Engineering careers, including SPE"s "*Members in Transition,*" which provides information, skills and tools to both new graduates and experienced professionals in oil and gas, disseminating best practices in career planning, job search, entrepreneurship and innovation. She has been an influential advocate for women's empowerment as a key member of SWE and advisor of SPE WIN (Women in Energy). She has also served on the Program Advisory Board for the Colorado School of Mines Petroleum Engineering Department, the External Advisory Committee and the Distinguished Alumni Committee for the Petroleum and Geosystems Engineering Department of UT Austin and on the Board of Directors of Girls Scouts—San Jacinto Council. Susan is a tireless leader, constantly planting seeds with the unselfish goal of helping others.

- Recognized industry leader in petro-technical talent attraction, development and retention. Expert in Reservoir Management, Risk Management and Portfolio Rationalization.
- Received the highest individual recognition of the Society of Petroleum Engineers, SPE, Honorary Membership in 2018.
- She has also received the SPE DeGolyer Distinguished Service Medal (2016), the International Distinguished Service Award (2005), the International Distinguished Member (2003) and the International Young Member Outstanding Service Award (1997)
- Susan has held numerous leadership roles in SPE.
- Renowned author, speaker and advocate of soft skills, knowledge management and strategic staffing for the energy sector with more than 30 articles

and more than 100 engagements as a speaker in international events. She is a SPE Distinguished Lecturer for 2019/2020 on the topic of Ethics.
- Currently Vice President of Engineering of Subsurface Consultants & Associates, LLC.
- Worked in Chevron and Anadarko, as Reservoir Management Organizational Capability Consultant, Reservoir Management Framework Consultant, Horizons Program Manager, Learning and Organizational Development Manager, Recruitment Manager, and Senior Staff Engineer
- B. Sc. In Petroleum Engineering, Cum Laude, from the University of Texas at Austin.
- A true Texan who loves bluebonnets.

A Personal Snapshot

Susan has been the go-to-person for key initiatives that we, Eve and Maria Angela, have advanced through the years. Eve met Susan in 1990 through SPE when Susan was working at Anadarko and Maria met her in Kuwait in 2009. Susan was one of the first people to recognize that professional societies such as SPE could and should play a major role in soft skills training as a charter member of the SPE Soft Skills Council. She has had a major role enhancing the awareness of the importance of mentoring, sponsoring and coaching.

The Interview

For the many organizations for which she is a volunteer, as well as for her employers who have had the benefit of her many contributions, Susan is an example of what tenacity, innovation and networking can do for an organization in numerous ways including reservoir management, onboarding of new hires, recruitment, and training.

Seeking Advice

Q. *Please tell us about your own mentoring experiences. Have you mentored many people? Do young professionals reach out to you to receive mentoring? How do your mentoring processes start?*

My mentoring happens at the office, at work, and when I volunteer at SPE. I cannot tell you how many people I have mentored. I began during my university years, when senior class students mentored the sophomores. When you realize you know a little bit more than someone else, it triggers mentoring. If someone else asks you something, and they receive that for which they are looking, they come back for more.

Q. *Do you like mentoring?*

I do! I have sought advice from people throughout my career and pass it on. For me, it is a pay it forward process. I want to make sure the insights and vision I have received are passed on.

Q. *Now, tell us about sponsoring. Have you been sponsored to advance in your career? Have you sponsored others during your careers?*

I would say yes, I have benefitted from sponsoring and I have sponsored people.

Sometimes you are not fully aware of how those discussions take place. It may become clear you have had help when someone who has been your sponsor moves out of the organization. What you may have enjoyed suddenly changes, and you need to find new sponsors in your role or new role. I try to do that as well.

Sponsoring is advocacy; it is endorsing someone for a role. Helping to do succession planning for organizations has been very interesting for me. In succession planning, you need to think about what experiences employees need to have before they ascend to top roles. Good sponsors in organizations are those who keep succession moving smoothly forward in an orderly fashion, providing opportunities to promising people.

Q. *When did you recognize that you needed to actively seek sponsors?*

Very early on, I realized sponsoring was important, even as early as when college students go into hiring processes for their entry job. At their starting point, they come in, but they notice that perhaps their peers are advancing more or less quickly, and they notice the impact of sponsoring. This is a clear case of "*whom you know*" opening doors and providing opportunities.

Most of us are interested in growth and challenges with opportunities to solve new problems. If you have sponsors, you are asked to join teams and given opportunities to advance.

Q. How far were you in your career when you realized you needed sponsors.

It was around when I had completed five years at work. It was clear to me that in the company for which I was working, if I asked for opportunities and I got the endorsement of key players, I could get those chances. It was not waiting to be noticed. You must be able to see yourself out in a role doing something you had not done before.

One of the very early opportunities I received was reviewing the technical translation from English to French of a 200-page long contract. This was clearly not my core discipline, but I was involved in the team because I expressed interest in the project. At the end, it opened other doors.

Q. Do you have a sense of how many sponsors you needed? How do you approach seeking out for sponsors?

For me, it is not about quantity, it is about having the right person at the right time. A sponsor needs to be able to lend credibility to you. If they are not in the group of people who are making the decisions, it is plainly not useful. Again, right people at the right time in the right place.

Q. How intense is your sponsoring?

I start by analyzing the roles at stake. It helps to understand the role for which someone is being considered and the key attributes that are needed for the role.

I based some decisions about sponsoring on behavioral-based interviewing techniques. I looked at the past performance of individuals to identify specific examples that could contribute to success in the open role. I look for "evidence."

Early in my career at Anadarko, I was involved in campus recruitment teams hiring fresh graduates. I learned how best to compare students, because many had similar CVs and internships. I learned what to look for as part of their interview process. A lot of it is just asking the right questions and talking to people. Once you learn about their experiences, you naturally discover how to advocate and be a sponsor for the best candidate.

Q. And that triggers a natural question: What happens when someone you sponsor does not perform?

Whenever you work with people, there are always examples of success and lack of success. That goes with the territory. You are going to have unexpected outcomes. For the most part, you work with it, using any opportunities to gather information to reduce the uncertainties and mitigate the risk of not knowing the outcomes.

Q. In your own experience, what are the questions that the mentees should ask you?

I think mentoring works best when the mentor is not driving the agenda, but rather when the mentoring experience is up to the mentee to ask questions or come to the mentor with a problem. It is useful when the mentee explains what they have tried and provides a description of what they are trying to solve. When they have a topic to be discussed, sometimes the process of defining the topic prompts the mentees to do a little research of materials that could be useful in defining the agenda. That is helpful and they start to learn to be self-reliant.

The best mentors are those who really challenge mentees to think. "*What have you already tried …?*" They ask open-ended questions, so that the protégé does not just follow procedures but uses critical thinking.

Q. Do you think that single sessions, or mini-mentoring sessions are also effective?

Chance meetings, an over-a-coffee or similar encounter, can transform into an effective mentoring session. Many people do not really appreciate the value you can retrieve from chance encounters.

I had a person at the energy workshop say, "Susan, do you remember, at OTC, when you asked if I would consider moving to Houston? Well, I did. I moved to Houston and it was good for my career!" I did not tell her to move to Houston. I posed the question for her own assessment of opportunities.

Short encounters must be very focused to single issues but can have big impact.

I am sure Eve has had many interactions like these short ones while SPE president.

(Eve answers): "Yes, it is surprising the influence you can have on someone's life with just a short chat."

Q. Do you think mentoring must be in person, or are on-line, digital mentoring solutions equally effective?

I have had several digital or virtual mentoring sessions and would say the mentoring experience is always enhanced face-to-face.

Q.Some people had a special influence in our careers. Mentors and most particularly sponsors are some of those influential people. Please tell us about your own experience as mentee. Who mentored you in a memorable way? Who sponsored you?

I have had the support of several mentors and sponsors throughout my career. Early on these senior engineers were at Anadarko. They were helping me develop critical skills, rather than giving me the answers. To have a mentor who helps you learn to think is the best kind of mentor you can have.

At SPE as I moved into leadership roles, I had a remarkable number of mentors and sponsors—a long list! That is the beauty of being involved in SPE, you learn different ways of doing things.

Sometimes you get contrasting advice from people outside your organization. In your company, often there is only one way to solve or tackle a problem. Seeing other ways from other companies, countries or regions has been useful.

Q. Do you remember a special one?

Yes. Early on in my career at Anadarko, I was involved in acquisition and divestitures. I was given the responsibility for a project to divest assets that was called "Momentum." It was a first for the company! I learned what to do by reaching outside my company, because no one had experience internally. I got a lot of guidance and support from a variety of people in the industry on coordinating data rooms and negotiating domestic and international asset sales.

Q. Is gender or age important in mentoring and in sponsoring? Should females be mentoring females or should more-experienced professionals mentor younger people?

I think the best mentor is someone who has at least one characteristic different than your own, because you need to frame and to see things differently than you do. Looking for someone who has a different technical discipline than yours, a different alma mater, different strengths … If your mentor is someone exactly like you, they will not challenge you and will not be valuable for your growth. Even a mentor who is younger than you can be useful, because reverse mentoring happens. Also, I think it is valuable to have men mentoring women, and women mentoring men.

Q. What about cultural challenges in mentoring?

Mentoring means different things for different people. In Chevron's "Horizons" program, a global program for the onboarding of technical new hires, the mentor should not also be the supervisor. Some of the Asian supervisors wanted to also be the mentor for their new employees. In Asian cultures, a mentor is traditionally a person with 25 years or more of experience, so that limits the number of mentors. In western cultures, a few more years of experience would be enough to make someone a mentor of less-experienced professionals.

In Asia Pacific, we had employees in a high role (labeled "Chief") as mentors, to accommodate the 25 years threshold expected in that culture for mentors. In other regions of the world, you may have similar situations not only because of age but also gender. For example, I received alerts like, "*My wife does not want me to mentor women.*" So, instead of pairing men or women randomly, you had to take into consideration these cultural nuances and adapt your plan to the cultural settings.

Other cases involved the problem of having very few mentors, leading to fulltime mentors who would have to deal with several protégés. This situation sometimes results from cultural expectations. In several cases, men did not want or simply could not mentor women in a one-on-one situation. You need to respect the societal framework to make mentoring work.

Q. *What happens is there is a misunderstanding?*

Different arrangements should be considered. For many cultures, mentoring is not just done, or it must be implemented in larger groups of people.

Q. *We have seen mentoring programs built blindly in Excel, assigning mentors to mentees almost blindly. It seems almost as if those were implemented to "tick" a Key Performance Indicator (KPI) goal of the department. What is your opinion of this practice?*

I have never been in a situation where mentoring is done blindly. In many cases, in which the "mentoring" was a form of technical coaching, I have seen mentors assigned by discipline, reservoirs engineers for reservoir engineers, for example.

Q. *What do you wish you did differently regarding mentoring or sponsoring, as a mentor or as a mentee?*

When I was managing the Horizons program, I was responsible for training mentors and for tracking mentoring metrics for the new hires. In hindsight, I would have made sure the assigned mentors were providing good value. In some cases, the mentors were excellent, but in others, not. Formal mentoring programs require more oversight than voluntary arrangements where people opt to join. I spent a lot of time trying to make sure that everyone had the resources needed for successful mentoring. I hope I did enough.

Q. *Do you think you can teach how to mentor?*

You can provide some skills to people that are useful for mentoring. You may empower them with tools that they can use to be better mentors. This is particularly useful for many people who were not natural mentors. It requires some organizational planning to make sure mentors are not overwhelming the protégés with too much information.

There is a book by Steve Trautman entitled, "*Teach What You Know: A Practical Leader's Guide to Knowledge Transfer Using Peer Mentoring*," that describes onboarding with an analogy to Maslow's hierarchy of needs: You need a PC, you need a password to log on, you need to know where the restrooms are. You need some basic things to get started in the corporate ecosystem. After that, a mentor should help a protégé to understand the team's mission, the deliverables, the deadlines and the big picture items. The mentor can then help programmatically to address the skills that are needed for the protégés to accomplish their work. We can teach the mentors systematic approaches if mentoring does not come naturally to them.

Q. *How do you measure success in mentoring?*

In some instances, the measure of success is the time to promotion, but in early career, there is not that much variation in the first promotion. Take for instance, a core group of peers who started work together. They will typically get their first promotion in their area or discipline at the same time. As they advance, some become the shining stars and that shows. These stars are getting more opportunities and new assignments than their peer group. The success of a good mentor can be demonstrated by the achievements of a protégé.

Alternatively, for example, when implementation of a project gets approved as recommended by a specific person, who has been mentored, there are signals that enable you to measure the success of mentoring in very concrete ways.

Are they getting good support from management? Do they prioritize appropriately? Those things may be harder to measure, but it could be valuable to assess the efficacy of mentors based on the success of their protégés.

Q. *Did you enjoy any mentoring or sponsoring experience?*

I have had numerous good experiences as a mentor and as a protégé. All of us are products of our experiences. I value very highly connections that I have made in my career. I still connect with them and aim to stay in touch.

(Eve comments) *When I was with Mobil, in the late 1980s and 1990s, there were many rounds of downsizing. Many managers lost their power and their groups were decimated more than others. Organizations grow or change dependent on who is at the top.*

Yes, change in organizations is so dependent on retaining expertise. I would like to highlight something I consider important for mentees: Don't ask "Will you be my mentor?" Instead, ask about a specific problem. Something you want to solve. I can't think of a time when I did not ask a question. For me, that is the natural approach.

Impression of Susan

Maria Angela crafted her impression of Susan.

- **Favorite role models?** Susan has so many contacts, she didn't single out a single role model for our interview.
- **Favorite way of mentoring?** Susan is a true expert in mentoring and uses a wide range of approaches. In person and with great efficiency via WhatsApp, phone calls and classic email.
- **Her connection with SPE?** A fair statement would be to say that Susan is almost synonymous with SPE.
- **What did she do for our book?** Susan was the guest editor of our book.
- Her name? Her full name is Carol Susan, but everybody knows her as "Susan," except her two daughters, Rachel and Elizabeth, who call her "Mom".

Afterthought

This was most probably one of the most difficult and at the same time easiest of the interviews we have included in our book.

Difficult and easy at the same time, because Susan is well known to both of us, and because she was our guest editor. Surely, her insights were needed, in any story that we wrote about mentoring and sponsoring, because she has been both, a mentor and a sponsor for us.

Alvaro Celis

"Causes need the right sponsors"

M. A. Capello and E. Sprunt, *Mentoring and Sponsoring*,
https://doi.org/10.1007/978-3-030-59433-6_19

A Glimpse

Reaching an impressive leadership position in one of the largest ecosystems in the technology industry is nothing less than remarkable, and Alvaro Celis has done it. He is the Vice President of Worldwide Device Partner Solution Sales at Microsoft and has global responsibility for a key segment of the corporation.

He graciously agreed to be interviewed about his career with the understanding that he was speaking as an individual and not as a representative of his company. Celis shared his reflections, personal insights and provided perceptive anecdotes on mentoring and sponsoring based on his multicultural experiences. His career has taken him from Venezuela to Brazil to Singapore and now to the USA. As his responsibilities grew, he became an outstanding leader, mentor, and sponsor for many others.

He inspires numerous people within and outside his organization, as he has made paying back to the community and to his roots a focus of his life. His three guiding principles in life are Family, Integrity and Passion, and that shows in his work for two initiatives he has embraced and committed to help boost Latino[1] minorities in technology: "*HOLA*", and "*HITEC*".

- Microsoft Vice President of Worldwide Device Partner Solution Sales, managing OEMs and Commercial Channel partners around the world.
- Alvaro has worked for Microsoft since 1992. Previous roles include Vice President of Sales, Marketing, Services, Information Technology, and Operations Group for Microsoft Asia–Pacific, Regional Operations for Microsoft's Multi Country Americas region, General Manager of Microsoft's Venezuela subsidiary, and in Guatemala and El Salvador.
- Global conference speaker on enterprise competitiveness, cloud, mobility, big data, and the empowerment of Latinx minorities in technology.
- Alvaro has been awarded the 2019 and 2020 HITEC[2] 100 awards for his influential and notable contributions to Latinx professionals in the technology sector.

[1]In USA, Latino and Hispanic refer to the inhabitants of the US who are of Latin American or Spanish origin. The term "Latinx" is often used instead of Latino and Latina, to incorporate gender fluid or non-binary individuals of Latin-American origin. We are using both terms in this chapter.

[2]HITEC is a global executive leadership organization of senior business and technology executives who have built outstanding careers in technology leading the largest Hispanic-owned technology firms across the Americas.

- Executive Sponsor for HOLA,[3] the Hispanic and LatinX employee resource group at Microsoft.
- Computer Science Engineer—Cum Laude, Universidad Simón Bolívar, Venezuela.

A Personal Snapshot

Alvaro immediately comes across as a straightforward person, very spontaneous, who openly shares his thoughts. He gains your trust from the first moment. We recognized that he has spent a great deal of time reflecting about his own role in leading initiatives for Hispanic/Latinx professionals in technology and on the importance of sponsoring people to accelerate careers in technology roles.

His gratitude to his sponsors speaks highly of his awareness of the mentors and sponsors who helped shape and propel his career. He is obviously focused on paying it forward and is actively engaged in a personal quest to support others to succeed in their careers.

The Interview

We conducted this interview during the reopening phase of the 2020 pandemic lockdown. Alvaro Celis graciously took time to speak to us while he was responding to the requests of companies to update and refresh their technology platforms to better cope with the requirements of remote and virtual work. Our interview was conducted remotely.

Family, Integrity, and Passion

Q. Alvaro, please tell us about your own mentoring and sponsoring experiences. Did any mentor or sponsor play a crucial role in your career?

I am doing this interview at a personal level. Not representing Microsoft. A human being sharing a personal journey.

[3] HOLA (Hispanic and Latinx Organization of Leaders in Action), is Microsoft's Hispanic/Latinx Employee Resource Group.

I want to share with you that I come from a very humble family in Venezuela. I never dreamed I would be managing a significant business worldwide. If someone would have told me this was going to be my future, I would have laughed, as I would not have believed this was feasible.

And yet, here I am! What happened is through my career, I have always had empathy for all. I truly believe in three cornerstone values: family, integrity and passion.

For my "family" value, I create for others a workplace environment with respect that enables everyone to project their authentic self and have a sense of family.

I believe in "integrity," and I do not like politics. I like an environment of trust with an ethics contract of how we work with each other.

I learned "passion" at work and in life from my mother. She died at 72 years of age, but we, her two kids, learned from her about having passion for living. She had her own business and was a propeller of our family's sustainability and progress.

I learned that your word is your vow.

Engaging in mentoring and sponsoring is not something I was taught to do, nor trained to do. I have an instinct for how I can help others. I am grateful for what I have achieved and look for how I can pay it forward.

It was my natural response to do something about poverty, minorities, and injustice. How can I pass the gift to others? Do you have it in your own heart? Believing in others is important.

I believe the more you understand the dynamics, the more you realize how you can help others. Therefore, I joined HITEC, which is an organization created to help Latinx in technology. Later, I engaged as exec sponsor for an employee resource group to support Hispanic/Latinx within my corporation, called "HOLA".

I am motivated by the question, "What can I do to pay it forward and help people to achieve all that they can be?"

Q. *You launched specific opportunities for Latinos and Latinas in technology. How was this initiated?*

The more I understand the dynamics of the Latino communities in the US, generally called Hispanic, Latinos and Latinas or Latinx, the more I understand how important it is to embrace your true self and not disguise it. Unfortunately, a vast majority of Latinx apply a lot of energy to disguising their true self.

HOLA is self-organized by employees. I have been with Microsoft for a long time—19 years in Latin America, then in Asia, and now in the US for a total of 28 years. I started to see and learn that 75% of Latinx in North America feel they need to hide their true self in what is called "covering." The more I understood the situation, the more I had to do something about it. So, I became sponsor of the group, responsible for helping them progress in the company and support the broader community.

There was a pivotal moment that occurred when the company appointed a new CEO. He was interested in creating a culture of inclusion within the company. This revolutionized the way the whole company looked at these topics. They created connections as an advisory group. I was fortunate in the timing as I was there. I am Latino and could engage and collaborate with the leadership of our company.

There was of course a business mindset about inclusion goals, with specific targets and measurements of progress. But for me, it is about employee bonds. To have a network that can support them. This can help provide the confidence to be yourself and make a difference for others.

Q. *In your opinion, what are the main hurdles experienced by Latinx in the US?*

I clearly see several hurdles, and probably the most important is not being your true self. The second one is to believe that if you do great things, great things will come to you automatically if you work hard.

As you progress in your leadership accountability in any company, achieving success takes more than hard work. You need to know how to influence and recruit partners and sponsors. How do we complement our technical skills and our leadership influence with more planning about our careers? We need to find sponsors who can open doors and provide opportunities.

Q. *From your experience, are multiple sponsors are needed?*

Absolutely! You must have many sponsors. A variety of sponsors inside and even outside your organization. In my case, I didn't recruit them. They chose me. I didn't know sponsors were needed and that they were important. I realized it when they moved on.

Huge endorsement is needed to tilt the balance in your favor. You must tap into their network.

Q. *And how do you select whom to sponsor?*

Having an eye for talent is important. For example, if you know the person, you understand if she is fit for specific roles. You can vouch for talented individuals, who have demonstrated their value. You must know your people.

Q. *Is there any risk in sponsoring?*

Sponsors take risks. They select who to sponsor, so it is important to recognize not only the capabilities of the individuals as demonstrated with technical accomplishments, but also their hunger to succeed, which is more difficult is to detect.

In most cases, I wasn't the person who was ready for the job, but my sponsors knew I wasn't afraid to learn fast and knew that at the end I would be successful. I have that hunger and raw talent, and they saw potential.

The sponsoring I received fueled my fire to move forward because I knew they took a risk on me.

Q. *Are there any sponsors you remember as pivotal in your career?*

There have been a couple of sponsors I remember with admiration. One was the Vice President of Latin America. He pulled me out of my country when I was 33 years old. He recognized my cycle of learning and contributions was fulfilled as a country general manager. He saw I did not want to move, and he pushed me super hard, suggesting that to make a difference, I needed to leave Venezuela and go to work in other countries and regions. That it would enable me to keep developing my leadership skills. He took me out of my comfort zone. It was a difficult battle with myself to leave my country behind and aim higher. I can assure you I left Venezuela "kicking and screaming."

He knew what was best for me. He vouched for me. Later, he was also the sponsor who pulled me out of the Latin-American region and sent me to a lead role in another region.

Q. *Is he still your sponsor?*

No, he retired, but he is still a good friend, coach and mentor. I am still connected to him.

All my sponsors knew I was not afraid and had resilience and determination to progress in my career.

Q. *Do you need seven sponsors? One of the people we interviewed mentioned seven as the magic number of sponsors for a successful career.*

I do not know if such a magic number exists, but you definitively need more than one. You need sponsors inside and outside of your company.

Q. *How did you embrace being passionate about helping Hispanic, Latinos or Latinas, Latinx?*

When I look in the mirror, do I feel good about it? Would my kids be proud? Am I doing something that will last after I'm gone? These are questions

I constantly ask myself. My quest and my cause are to achieve equality. I want to be sure we level the playing field and remove the biases that cause exclusion. That's where you win. Bring a more factual, egalitarian approach.

I ask myself; how can you help by sponsoring and providing adequate support to the people who need that to give them a real chance to compete? If you do it too aggressively, you just harm the cause, because people fail, and it takes away the credibility from everybody.

I always try to reflect on my portfolio of sponsored people, to make sure I'm fighting my own biases. I have in my portfolio women, peers, Americans, Asians, early career and experienced people as well as people inside and outside the company for which I work.

I have a daughter and that gives me a sensibility to the gender inclusion, because I want her to have a full chance. My youngest is my son who is on the autism spectrum. He is a high functioning kid, but how do I help him? I try to be supportive to the neuro-diversity cause. Am I the right person to move it forward? Are you the right sponsor or can you help that person get the right sponsor?

There are people who are very early in their career whom I sponsor and support. As they are the right leaders for the challenge ahead.

Q. *Do you have an example of sponsoring for you or others from outside your work loops?*

I have a variety of sponsors. For example, I had a sponsor external to my company, who took me to HITEC. He was the one who put me on the fast track when he saw my passion. I connected, and he helped me realize that path of service to the community.

Working with minorities is a complex matter. Sometimes, there is a radical element. I am a proud Hispanic, Latino, Latinx. In that sense, my quest, my cause is a cause for equality, for Latinx to be recognized for their merits without biases, for what they deserve. We want our people to be the best they can be without exclusion from the system.

I wish to be a person who tries to level the playing field and leads the conversation for a more equalitarian approach. Human processes are intrinsically complex. Opportunities in a career typically have criteria with a technical component, as well as a soft skills or behavioral component.

What is effective for sponsoring minorities is bifold: (a) for the people to be ready to be considered for promotions or changes, and (b) for mechanisms to be in place for getting more people from minorities to be in that position.

If you start to sponsor too aggressively, and push people when they are not ready, it will completely take away the credibility of anyone else. We don't want to create single or solitary lighthouses.

Q. *Since sponsoring by its very nature is not an equitable process and is dependent on what someone sees in someone else, do you find yourself sponsoring in an unbiased way?*

I am more attracted to a cause, than to a person.

Sometimes, you are the right person to move a cause forward. Only then you can be a sponsor. Causes need the right sponsor, a person who represents the cause.

I try to be as unbiased as possible, but I am human. The challenging thing for me is when to say no.

Q. *You are truly immersed in sponsoring. How many people have you sponsored?*

I have sponsored quite a few both inside and outside the company, probably hundreds.

Q. *In your own experience, what are the questions that the mentees should ask in a mentoring session?*

First, it is important to determine if they are seeking a technical solution, in which case they might need a tech expert. Then, it is extremely important for the mentee to explain some frames. Are the mentees serious? Usually they ask me what I expect from the relationship. Then, respect for my time and their time should be part of the mentoring scheme.

I like those who come with questions about the questions. For example, *Am I thinking about this in the right way?* If the person, the mentee, is really trying to sharpen their thought process, you need to help them find their own answers.

Mentoring is about supporting and enriching thought processes. Not about telling them what to do.

Q. *Which do you prefer, mentoring or sponsoring?*

I have a soft spot for sponsoring. You see your sponsored people growing up and moving up through levels. It is satisfying. You are more vested in sponsoring, so there is a different level of emotional satisfaction. When you see the blooming, that is superb!

With mentoring even if you still want them to break through, you haven't invested your personal equity or reputation.

Q. *Do you think that single sessions, or mini-mentoring sessions are also effective?*

I am more outcome related. Can you help someone think differently about something? I am motivated by this aim in mentoring. When someone is not familiar with the process it takes longer.

Q. *Do you think mentoring has to be in person, or are on-line, digital mentoring solutions equally effective? This is a funny question, for someone working at Microsoft!*

I say you can make it work. Look at our conversation right now. You can still look at people in the eyes. You can make it work perfectly. If I have the chance to sit down face-to-face, it is super powerful.

Is it needed? No. Will it help bootstrap the chemistry? Yes.

The best thing is to do a mix. Personal contact to establish a rapport and then carry on with a hybrid approach.

Q. *Some people have a special influence in our careers. Mentors and most particularly sponsors are some of those influential people. Please tell us about your own experience as mentee. Any anecdotes?*

I was in Asia–Pacific and one of my mentors helped me think through a piece of feedback on what I needed to do to move forward. I want to share this, as it was a memorable experience of my own mentoring.

My first reaction was defensive. I didn't get it. "*Why me?*" I was puzzled.

My mentor asked me a couple of good questions including, "*Why do you think that?*" and "*Imagine yourself in this situation. What would you do?*"

I realized that I was thinking about it all wrong. The feedback hadn't sounded genuine to me. He helped me think through why I thought I was getting this feedback. Through his questions, I discovered a platform that took me to a higher level of leadership. He helped me find my answer. He didn't tell me what to do. Those are the mentoring relationships that are transformative.

Q. *Is gender or age important in mentoring? And in sponsoring?*

I don't think so. Another one of my mentors in Asia Pacific was a very early career Asian female. It was a super strong opportunity for me to grow and learn. I discovered new ways to look at things. She was a fantastic mentor for me, and it was an extraordinary case of what we call today "reverse" mentoring.

You need to have the right mindset. You need to have the right attitude.

Q. What do you wish you would have done differently in reference to mentoring or sponsoring?

I could have been more structured. I had for a long time a very natural and ad-hoc approach to mentoring and sponsoring. If I had more structure and discipline earlier, I would have been able to help more people and had more impact.

A Shared Selfie

- **Your favorite role model**: You always admire people. I admire Bill Gates, Tim Cook, Jack Welch, iconic leaders. Also, Orlando Ayala one of the most passionate leaders for whom I ever worked, and Eugenio Beaufrand, Microsoft VP of Latin America.
- **One word to define your experience with mentoring**: Transformation.
- **One word to define your experience with sponsoring**: Commitment.
- **Who did you want to be your mentor/sponsor, but you never had the chance to ask**? Probably Jack Welch, former CEO of GE. When you look at the story, he was such a remarkable transformational force. Picking his brain would be awesome.
- **Who would be an ideal mentor for you in this moment of your life?** Simon Bolivar. As a Venezuelan, this is my choice!
- **What question would you ask this mentor?** How did he get the courage and ambition to change the status quo of the South American continent and give up his fortune to do it and overcome the real challenges? That mentor is close to my roots, close to my history. I'm amazed by his energy and inspiration to take those big risks and leave behind his advantages.
- **What question would you like Simon Bolivar to ask you?** Who are you and what do you want?

Afterthought

As a thought exercise, at the end of the interview we asked Alvaro the same question he imagined Simon Bolivar would ask him. "*What do you want, Alvaro?*"

His answer was awe-inspiring: "*I want to help to make a difference.*"

We are convinced that HOLA and HITEC have in Alvaro Celis an extraordinary role model, who is pushing to make a difference. He has realized the importance of supporting Latinx progression in technology.

It was refreshing to speak with such a down to earth, pragmatic person, who retains idealism in his life. It was wonderful to see his genuine emotion, when we challenged him to imagine he was in a time-machine and could speak to Simon Bolivar.

Dr. Wafik Beydoun

"I do not see mentoring as an extra activity. I see it as a natural conversation."

M. A. Capello and E. Sprunt, *Mentoring and Sponsoring*,
https://doi.org/10.1007/978-3-030-59433-6_20

A Glimpse

We were very glad to include Dr. Wafik Beydoun in our compilation of leaders with insights on mentoring and sponsoring, especially because he is a natural sponsor. We have seen him in action, promoting and advocating for young and experienced professionals, reaching outside his department, company or even the region in which he was working, to support and express a good word for someone aspiring to a promotion, or looking for a new job.

With a truly multicultural upbringing and career journey, the gentle and at the same time decisive manners of Dr. Beydoun have brought the respect of peers and colleagues all over the world. He has been a very dedicated and active volunteer with leading roles in several professional societies, and most recently, was the chair of the most important technical conference and exhibition in oil and gas: the Offshore Technology Conference, a giant networking and knowledge sharing event held annually in Houston TX, USA.

- Director for the Americas at the International Association of Oil and Gas Producers (IOGP).
- Chairman of the Board, Offshore Technology Conference (OTC) 2018–2019.
- President and CEO, Total E&P Research and Technology USA.
- Manager, R&D Division ADNOC UAE.
- Led numerous key programs, country businesses and initiatives for the giant French multinational operator Total, including Chair of TOTAL Kuwait, Business Development Manager Technology and R&D in Total France, Senior Negotiator New Ventures Total France, Manager of Geophysical Operations and Technology, Geosciences Research Centre, and Area Exploration Manager in Angola.
- Lived in 12 different countries on 4 continents, acquiring multicultural exposure that propels his insights on leadership and how mentoring needs to adapt to each setting.
- Board Member, Research Partnership to Secure Energy for America (RPSEA) 2012–2014.
- Vice President, Society of Exploration Geophysicists (SEG) 2011–2012.
- The American Association of Petroleum Geologists, AAPG, 2000 George C. Matson Award for Best oral paper "Exploration Challenges Into Angolan Deep Water to Ultra Deep Waters."
- M.Sc and Ph.D. in Earth Science/Geophysics from MIT, and Maitrise in Physics and Geophysics, Polytech Paris-UPMC.

A Personal Snapshot

We were very glad to include Dr. Wafik Beydoun in our compilation of leaders with insights on mentoring and sponsoring, especially because he is a natural sponsor. We have seen him in action, promoting and advocating for young and experienced professionals, reaching outside his department, company or even the region in which he was working, to support and express a good word for someone aspiring a promotion, or looking for a new job.

With a truly multicultural upbringing and career journey, the gentle and at the same time decisive manners of Dr. Beydoun have brought the respect of peers and colleagues all over the world. He has been a very dedicated and active volunteer with leading roles in several professional societies, and most recently, was the chair of the most important technical conference and exhibition in oil and gas: the Offshore Technology Conference, a giant networking and knowledge sharing event held annually in Houston, Texas, in the USA.

The Interview

Q. Please tell us about your own mentoring and sponsoring experiences.

I would use more the word "advise" instead of mentoring, as I find it more informal on behalf of both persons involved. Mentoring seems a very formal and official word to me. Regarding sponsoring, I tend to relate this word with financial or supporting activities. Because of this, I consider myself giving some advice when mentoring a mentee and associate the word "sponsoring" to supporting an individual in the course of his evolution in some organization.

In general, I have given advice in my career when requested. My position is, "I am not here to spread my experience," as there are people who freely and unsolicited, offer their advice and recommendations to colleagues or individuals in their teams. Even if I am leading a team, my preference on this matter is to have the person ask for advice. Only then, will I try addressing her or his request. And this approach applies to young and not so young professionals. Some of them ask, and some do not ask for advice or guidance. What has been very interesting for me is that the mentee's approach varies depending on the individual's background, culture, or if man or woman.

I do not see mentoring as an extra activity; I see it as a natural conversation, part of the work loops.

Q. Do you think that choosing a good mentor is also a result of luck, chance?

We are not only the product of experiences we have assimilated through time, we are also the result of the random opportunities that appear to us and can be seized. You said "chance" and yes, we are in a random world and sometimes we see in mentoring the chance effect. But of course, we also have around wonderful people who grace us with their time, dedication, trust and support along the way. These people enabled us to reveal our true and unique potential. They naturally inspired me to do the same.

Mentors are not only found in the professional world. Some could be parents, relatives, and/or friends. There is a long and growing list of people to whom I am personally very grateful for their advice and support.

The best way to explain my point is with an analogy. Career advice or mentoring is like what you may gain when you attend a large-format conference and visit the exhibition stands. Let us assume here that the individual, the mentee, visits the exhibition hall with a big empty bag.

The mentors would be the people in the stands who are promoting and displaying on their tables pieces of puzzles. These pieces of puzzle encapsulate in some way the mentors' experience and knowledge. The mentees while visiting different stands, would discuss with mentors and pick up puzzle pieces that appear interesting to them and put these in their bag. They could then come back later to visit mentors (stands) and pick up more pieces.

But having a bag full of puzzle pieces does not mean that the mentee has benefited from the mentors' advice. Pieces may or may not fit together, be redundant, and/or not even be relevant. The point I am trying to make is that the mentee can get the maximum benefit of what is in his bag if she or he takes and analyzes each piece, and keeps those that 'resonate' with her/his aspirations and background, and fill in the gaps when composing her/his personalized version of a puzzle. We may find excellent mentors along our career path, even a few roles models, but if the information (pieces in the bag) is not adapted, transformed and integrated into a personalized guideline for the mentee to implement, then these puzzle pieces are just a patchwork of other people experiences.

It is a two-way conversation. If the mentee can formulate questions, drivers, and aspirations, then the mentor could add new pieces of puzzles on the table that are more relevant to the conversation.

Mentors would most certainly be willing to give back to the younger generation. In a way, senior mentors should have in their track record many jigsaw puzzles and many pieces, right? But here also, the young mentee needs to express, as clearly as he can, her/his aspirations. The overall mentor–mentee dialogue is not in my opinion an explicit and deterministic process. There is randomness imbedded in it.

Q. *You are saying that there is a random factor in how one advances a career. Now, tell us about sponsoring. Have you been sponsored to progress your career? You just helped me with a possible opportunity at a large conference, I consider you a natural sponsor.*

I do not consider myself a natural sponsor, maybe because I would like the process to be more direct and efficient. The effectiveness of sponsoring depends on the person's role in the organization, and her/his ability to make (or influence) the decisions.

I will share one of my own examples of sponsoring to illustrate this point. I spent some time at ADNOC, the national oil and gas operator of Abu Dhabi in the UAE, as a secondee of Total. Sponsoring people in my team was not straightforward in my role of secondee. There was this engineer—a wonderful person, technically and personally. When I tried to promote this person in the organization, it did not work the first time. At the end, after many attempts and much later, she finally got a big, well deserved promotion. My 'sponsoring' may have worked, but it took much longer than expected. I must admit my mentee was very resilient, although out of frustration, she even thought about leaving the company. The message here is: don't stop trying but remain patient and determined. And I guess that, as a mentor and sponsor, we need to stay motivated and … keep trying!

Q. *What about the sponsoring loops in large corporations? Are those different than the loops in small companies?*

The larger the organization, the more you need to have facilitators or sponsors. The governance being more complex, there are generally HR processes that evaluate and promote high potential individuals. People in large organizations have then less interest in helping a specific individual. But understanding your network of influences in this case and identifying those who may have interest in you and your potential, is quite important.

Is it a win–win situation? That depends on the size of the organization. In large corporations such as in Shell, BP or Total, the organization is wide and deep, and you must observe and respect processes. This is logical, as there are many people who are managed, and compete sometimes for a same position. Many times it is a matter of remaining patient and engaged, but also remaining open for other 'equivalent' positions that may appear in some random manner.

It is not an easy process to climb the corporate ladder. You need to demonstrate how you contribute to the organization's objectives. More than once, more than twice. You must learn how to communicate your contributions to the management and have them 'connect the dots' on how you could have a sustaining positive impact in the corporation by making you grow in it.

Q. Have you been sponsored?

Yes, I have been sponsored. And these persons went through some hurdles to promote my work and abilities. It sometimes lead nowhere. But I remained engaged and patient. And then it happened.

Q. How do you keep your network of sponsors? Do you update your network of sponsors?

That is a tricky question. You need to respect the governance of your company. If your sponsor is above your direct supervisors, special care should be taken. I made sure my direct supervisors were involved in the process, when soliciting my sponsors who were two or three levels above them. You do not want to use that big leverage often. If I was interested about an opportunity, I would make sure to inform my supervisors, direct and above, so that no one would be bypassed in the exchange of information and especially in the decision making. Think of it reciprocally: Would you appreciate someone that you supervise be ing sponsored by one of your managers without you being involved? I would not.

People bypassing the chain of command demolishes the effectiveness of an organization.

The corporate sponsoring processes are not straight forward, and you must understand the dynamics. It is good to have sponsors, but you should not abuse that kind of privilege. Again, the most compelling element is your performance and results.

Q. In your own experience, what are the questions that the mentees should ask you?

The mentee needs to communicate to the mentor, even in some vague way, what he or she wants to do in some future—this is what is meant by "aspirations." This is the cornerstone, the element with which to start from. Remember the puzzle. The more the mentee can provide this type of information, the closer puzzle pieces provided by mentors would make sense, be put in the bag, analyzed and adapted.

But to answer your question, as a mentor, I would like to be asked by the mentee, "Could I first explain my aspirations and background, then please share your insights and advice?" I would then inquire if he or she is a start-up/entrepreneur person, interested in developing their experience in large or small organizations, etc. Alternatively, the mentee could also formulate open-ended questions related to her/his aspirations to kick-off the conversation.

As a counter example, I have had some prospective mentees asking me upfront, "Tell me what to do," or worse, "Can you introduce me to important people so that I can get-in or move up." Such questions are very demotivating for a mentor. As a mentee, you need a special mindset and some preparation before reaching out to mentors, and then to absorb these pieces of puzzle.

As a mentor, it would be condescending to assume that our experience alone would address the mentee's aspirations. You may have big puzzles on your table as an experienced person, but your vast experience may not fit in their own story, in their own puzzle, to make it right for them. Your mentee's background and aspiration are probably quite different than your own.

Q. *The puzzle analogy is so powerful! Did you have this puzzle idea now or long time ago?*

This idea came to me when I realized that no matter how much experience one may have, it was acquired and built in the course this person's specific (and likely unique) evolution and environment—which would certainly be different for another individual with different background and aspirations. You must then remain humble as a mentor. Personally, as a mentee, I needed many puzzle pieces in my bag to start building my own plan. It was not a single mentor who guided me throughout my career. There were several. At the end, you are the one who designs, adjusts and implement your own professional guideline, using some of puzzle pieces you have in your bag.

Q. *Do you consider that single sessions, or mini-mentoring sessions are also effective?*

It could be single sessions. Could be virtually, online, especially if you know well the person you are mentoring. But if I don't know the mentee, I prefer at least the first session to be in person.

After that connection is established, the mentoring could be effective using various channels.

Some people have had a special influence in our careers. Please tell us about your own. Were you mentored or sponsored by someone in a way?

Yes, and I have two experiences I would like to share, one about mentoring and the other about sponsoring.

When I got my baccalaureate in France, I aimed to study at the Paris VI Pierre and Marie Curie University in France. I was going to register in astrophysics.

Something happened on registration day. I was waiting in the registration line and there were some professors in the room to assist. I was aspiring to be an astrophysicist. I forgot who this person was, I don't remember the name, but I do remember very well the discussion. He asked me, "Why do you want to study astrophysics?" I answered that I liked understanding planets, acquiring data and being able to visualize, understand and explain observations through some models. I thought at the time that astrophysics had many theories, but few models. I was enthusiastic about acquiring data, seeking models that fit the data, and going beyond.

This professor listened with attention and then explained, "Astrophysics is an area where you have relatively limited data to discriminate among many competing theories/models. However, geophysics (being the astrophysics of planet Earth), you have thousands of data that can actually be used in a more practical way to validate earth models and also find earth resources (e.g., oil and gas, mining)." In summary, I was convinced by this idea, and this is what triggered me to venture in geophysics.

This is the randomness in destiny I was talking about earlier. An eighteen-year-old student changing his destiny in five minutes after a random encounter? It was a mesmerizing experience. Bottom line: Recognizing randomness made my aspiration more robust, enabling me to perceive and seize new opportunities.

It is important to remain open minded, and realize that this random factor is always there and may help you. You need to keep a space for randomness and remain flexible.

The second example is amusing and related to sponsorship. Once I got my French master's in physics and geophysics, I wanted to pursue graduate studies in the US, as at that time, leading edge geophysical research was conducted in the US. I applied to several universities in the US. Two women professors supported me in France, one in charge of Geophysics and the other one of Physics. It is thanks to their support and their recommendations that I was able to go the US and was admitted at MIT.

That kind of endorsement was key for me. But to complete the story, I also had a professor who knew me well, decline my request to write a recommendation letter to support my applications to US universities. "I will not support you," he told me—he was at least honest, but you can imagine my disappointment. Fortunately, the two women professors who supported me, were more engaged with the very concept of opportunity and encouraged me with a decisive "go ahead".

Q. *Is gender or age important in mentoring?*

I have been mentored by a younger person.

In France, the "Grandes Ecoles" (great engineering schools) provide the elite of engineers. They are generally hired by big companies, and follow a fast-track path towards executive positions. Age and what school your degree is from are important. Could these great engineers be better mentors? Not necessarily, as they also would tend to see mentees through their own lenses. I have also been in other situations in which age is important. In the Middle East, a young person mentoring an older one is not necessarily straightforward.

In a more general context, if the younger person is mentoring an older one, the mentee needs to be open minded. To get the most out of this encounter, he should focus on choosing the puzzle pieces offered by the younger person by removing the age-factor from the equation/perception.

Q. *What do you wish you had done differently in reference to mentoring or sponsoring, as a mentor or as a mentee?*

To accelerate the entire process.

All the sponsoring and mentoring I could have gained is now water under the bridge as 30+ years of career have passed, but I think people with experience can accelerate the process and support the younger generation, who need to be exposed to real-life situations as early as feasible.

I would say and advise my mentees to implement their ideas, learn from their mistakes and bounce back as many times as necessary, especially when they are young. If uncertain, to experiment with a few options fitting with their aspirations, maybe be exposed to different roles. We need to have organizations that enable young professionals to make mistakes. It is by failing that you learn and develop a more resilient path to success. I would have wished to accelerate this learning process.

Q. *Did you enjoy any mentoring or sponsoring experience?*

When I was a mentee, it was gratifying when my mentors believed in me. You feel obliged and somehow you want to share that pleasure with your mentor. I am now doing more mentoring.

I particularly enjoy seeing my mentees taking all the puzzle pieces and build their picture/story and implement it. The woman who I mentioned earlier, who was promoted after significant patience and endurance, was for her a big success, as well as for me—as a sponsor and mentor. I would like to believe that my mentoring effort paid off.

Sometimes, mentors are not aware of the positive influence they may have in other people's lives. They can have a tremendous impact.

Q. *Do you think social media are a new channel to create a new line of mentors? The influencers?*

People with interesting range of experience and a strong presence online may influence other people's careers and lives. You for example, Maria Angela, when you share your insights or experience, point of view in your Instagram, LinkedIn or Twitter account, you are placing your puzzle pieces into the hyperspace. People could pick the pieces they like, even without you knowing it. The person looking for guidance may recognize, in what you are sharing, a missing piece of their puzzle.

A Shared Selfie

- **Your favorite role model**: I have different ones, at different times or context. Let me confess that I still feel like a mentee even now. Even when I retire, I will have mentors. In academia I had several professors who were my role models, whose behaviors inspired me with a selfless attitude. They were enjoying giving back. I try to do this now. It is not a task or an obligation for me to mentor or sponsor others. When I see capable, curious and mature individuals, it comes naturally to try to support them, to give them advice. I know they will pay it forward.
- **One word to define your experience with mentoring**: Relationship.
- **One word to define your experience with sponsoring**: Nurturing.
- **Who would be an ideal mentor for you in this moment of your life?** They would be different people. If I had the opportunity to talk to historical figures, I think I would like Mustafa Kemal Ataturk to be my mentor. He was Turkey's leader who reunited Turkey after WWII. He saved Turkey after the Ottoman empire crumbled, preserving an identity while establishing a country. He is the founder of the modern Turkey. I would have been very interested to know how he came up with his salvation plan.
- **What would you ask Ataturk?** I would ask him, "What were the critical circumstances and challenges in your life and how did you transform those into an action plan?"

Afterthought

When we finalized the interview, Maria Angela told Wafik his choice of an ideal mentor was surprising, because he was raised in France, Brazil and Lebanon, and Turkey was not one of the more than a dozen countries where he has lived.

Wafik commented his choice was probably because of his enthusiasm for learning about decision making processes. He reads a lot about cases in which decisions had to be made to address extremely difficult situations. Ataturk was one of those about whom he really enjoyed reading. Sometimes, he said, mentors are not aware of the positive influence they may have in other people's lives. Some people think the grandeur of some leaders is destiny, but I think it may also be the result of randomness combined with a clever person, who knew how to use that random element and put it to work. That is an important ability.

Dr. Ana Gil García

"Latinas trust Latinas."

M. A. Capello and E. Sprunt, *Mentoring and Sponsoring*,
https://doi.org/10.1007/978-3-030-59433-6_21

A Glimpse

Dr. Ana Gil Garcia is a Venezuelan American professor. But that is a simplistic way to describe her. She is the powerhouse who champions strong advocacy for Hispanic Women, an unstoppable denouncer of barriers for gender equity, and an unrivalled promoter of the Fulbright ecosystem and democracy.

We have in Dr. Gil Garcia a truly fine role model for transforming ideals into actions. She is an Emeritus Professor of Northeastern Illinois University, a five-time Fulbright Scholar, an internationally acclaimed professional, a published author and an esteemed community leader and leading advocate for diversity.

Her involvement and advocacy have led her to take actions on the empowerment of Latina women through higher education, the promotion of coalitions among Hispanic professional educational organizations and the creation of international bridges for minority students. She enabled the launch of a principal endorsement program for Latino teachers, who will be the school leaders needed in Chicago Public Schools in which 49% of the student population are Latinos, by co-founding the Illinois Latino Coalition of Education Leaders (ILCEL). She is also the co-founder of the Inspiring Latinas, a not-for-profit organization that supports women in their 50s and beyond.

Her defense of democracy has brought recognition from diverse human rights groups. She is the co-founder of the Illinois Venezuelan Alliance, a not-for-profit organization to denounce the existing Venezuelan humanitarian crisis. Dr. Gil-Garcia's image served as one of the Rotary International faces of the "*Humanity in Action*" worldwide campaign to fight poverty.

- She is an expert in Educational Leadership: she holds an Education Doctorate from Western Michigan University, MSc from the University of Tennessee at Martin; an Ed. S. From Iinstituto Pedagogico de Maracay, and B. Sc from the Universidad Pedagogica Libertador.
- Awarded as one of the 100 most influential Hispanics in the USA (2010), one of the 30 Outstanding Visionary Women in higher education (2014), Distinguished Woman in Education (2015) and one of the 50 more influential Latinos in Chicagoland (2017). In 2018, the Illinois Association of Hispanic State Employees conferred upon her the Hilda Lopez Award. Recipient of the Latina Trailblazer Award, Civic Engagement Medal (Porlamar, Venezuela), and Faculty Excellence Award.
- Emeritus Professor of Northeastern Illinois University, and Visiting Professor in universities in Republic of Georgia, Saudi Arabia, Egypt, and

in the US, in Western Michigan University in Kalamazoo, Michigan and in Loyola University – Chicago.
- Author or co-author of 6 books and 30 articles in peer-reviewed journals; participated in 43 international presentations.
- Member, Director or Chair of the Illinois Venezuelan Alliance, Illinois Latino Coalition of Education Leaders, Latino Advisory Committee (Chicago Public Schools), LULAC Education, Council of Latino Affairs of the City of Chicago, Rotary Club of Chicago, Women's Business Club of University of Chicago, Chicago Fulbright Chapter, and many more, in an almost infinite list.
- Editorial Board Membership in 14 top peer-reviewed journals of research and practice of leadership in educaction
- Awarded Fulbright Scholar grants in Liberia, Georgia, Armenia, Venezuela, and the Middle Eastern countries of Qatar, Kuwait and UAE.
- Master of Art, Curriculum and Instruction, University of Tennessee at Martin, Educational Doctorate, Western Michigan University

A Personal Snapshot

Dr. Gil Garcia knows the world of education on a first-hand basis. She has travelled extensively in a career journey that has taken her to share her knowledge about education leadership from Chicago to Saudi Arabia. In one of her multiple trips, Dr. Ana Gil Garcia passed by Kuwait and it was there, back in 2011, when Maria met her. Dr. Gil Garcia's depth in terms of knowledge and human quality was obvious, and her charming personality sparkled brightly from the start of the liaison.

Through the years, the activity of Ana (as she wants people to call her, plain "Ana"), flourished incrementally. Elevating one's personal station is feasible, as is clearly demonstrated by Ana. She has collected funds to launch a library in her native town in Venezuela, and she has empowered Latinas with specific sustainable actions in Chicago. We wonder what is not feasible, if given to Ana.

The Interview

This was another of the interviews conducted from Rome, in what were indeed "Vacanze Romane," like the classic film. Eve was in California and Ana in Chicago, so the three of us were in different time zones, but the video conference worked well, and we hope our summary will reflect the enthusiasm that Ana is capable of transmitting in person.

Q. Please tell us about your own mentoring experiences.

I earned my first degree, a B.S. in Biology. I wanted to teach biology and chemistry. Then, I came to the US for a M.S. and gain more insights into the educational field. I then earned my Ph.D. in education and leadership. I am now planning to retire. I have had a very rich life, and I will not miss the classroom. I do not want to deal with the administrative side of teaching anymore.

Educational Leadership is my field. I worked in Chicago for 24 years. Worked a lot in the Latino community—which has a high deficit in terms of opportunities for females, who are a minority. I have been creating some programs for Latinos—especially for Latina females, to empower their way into academia. I have encouraged and even pushed them to consider the higher education path.

I would like to see more Latina professors.

So, to answer your question, yes, I have mentored. I have mentored many people directly and indirectly, young professionals reaching out about career paths, asking questions about their own process. I believe that my mentoring process has been a natural one. I do not have any formula. I usually approach someone with potential, for whom I can see a bright future. I then ask the person about her life plan. I push them to take graduate courses. (I only teach graduate courses). It flows in a very natural manner—nothing with any type of formula.

Mentoring is not part of what I do every day nor in my academic workload. It is more spontaneous. I have many of my own students and others outside the university contacting me for guidance. Therefore, I set up an appointment in a coffee shop, to create the connection and the relationship. Not a structured way of mentoring. The informality of mentoring is what makes it healthier.

Q. How many people have you mentored?

I wish I knew how many people I have mentored—I do not have that recorded. I have been 40 something years in my professional life. I wish I knew how many mentees I have positively influenced, where they went and what they did. I have mentored many people during my professional life. In fact, mostly, young professionals reach out to me to receive guidance, mentoring, advice, and to consult on career paths like mine.

Q. *How do you start your mentoring sessions?*

My mentoring process is a natural one. I do not have a formula. I have found that I am able to identify the personal potential of individuals whom I approach. I immediately ask that person what is their professional life plan. I insist that younger minority professionals consider pursuing their doctoral degrees due to the extreme deficit of minorities in that area.

Any critical question I ask triggers the mentoring session. I have found there are more questions to me than from me to them. They ask, for example "Do you think I can do this? What do you see in me that makes you think I can do this?" or "May I call you anytime?" and the very famous: "Do you think I'm ready for this job?"

When they call, I ask them whether they have met me before? Often, they just know me by a reference. I also ask whether they are interested in academia. I always wonder if they want to be a university professor. Then I tell them about my own career path. How I started in academia, what kinds of obstacles I faced, and how I overcame those obstacles.

Relationship is the basis.

Q. *Now, tell us about sponsoring. Have you been sponsored to advance in your career? How? Have you sponsored others during your careers?*

Yes, I have been sponsored several times to advance in my career. My sponsors have identified qualities in not only my personality but my professional dedication, integrity and discipline that made them think that I was the right person to pursue a certain path, for example a higher education career as an academician and or a professional developer.

I have sponsored others during my career. I am normally looking for opportunities for other people whom I believe may have what it takes to become a unique professional, especially in the field of educational leadership.

Q. *Do you remember any fun or challenging moments in your sponsoring processes or when you were sponsored?*

I would like to share three pivotal moments in my journey. Yes, you may call these three anecdotes, related to my own sponsors.

Dr. Eugene Thompson, the chairperson of the Educational Leadership department at Western Michigan University (WMU), my Alma Mater for my doctoral program, insisted that I needed to check "how marketable I was" in the US higher education academic market. I told him that I was not interested, that I had my own job in Venezuela and that as soon as I completed my doctoral work, I was going back to my faculty position at the Pedagogical University in Caracas. Almost every day for 10 days, he would open my office door at WMU, and place on my desk, clips from the Chronicle of Higher Education newspaper with different higher education job postings for which he thought I would be a good fit.

A week before departing for my home country, I decided to do something about it, if only to satisfy Dr. Thompson's personal ideas about my professional potential. So, I picked the advertised job postings, selected nine positions at different US universities, wrote letters of interest and I departed for Venezuela. Surprisingly, a week later, I was receiving emails, certified letters and phone calls from six higher education institutions inviting me for an interview at their campuses!!! I called Dr. Thompson to tell him what happened, and he said, You did not want to believe that you are "a rare individual for most universities here in US, a minority, because you are a Latina woman with a doctoral degree. What I mean is that you are a luxury for many institutions."

The second anecdote I have is with someone I learned to love as a professional, colleague, friend, confidant and sister, Dr. Ernestine Riggs. When I arrived in Chicago 24 years ago, I accepted the Assistant Professor position at Northeastern Illinois University, a state institution with the lowest salary among all state universities in Illinois. However, with the most diverse people, faculty, staff and students that made me feel like home, family. Because my salary was so low, and I had two children to feed, I needed to look for a second job to make some extra money. If I spent money for a coffee, for example, my personal budget would suffer. My department chairperson knew someone who was looking for a trainer, a consultant, a professional developer for an education project to be implemented in schools in Chicago. It was Dr. Ernestine Riggs, a pioneer school administrator, an African American woman, with strong reputation in the school system.

On April 17, 1996, four months after my arrival in Chicago, I received a phone call and she said, "Are you Ana Gil Garcia?" "Yes," I said. "I am calling on behalf of Dr. Jane Baxter. What are you doing tomorrow morning?" I said, "I do not have anything scheduled." She responded, "Here is an address and I would like to see you at 8:30 am at this elementary school." I said, "I will be there." I arrived on time and I was asked to go to the school auditorium for a meeting. When I stepped in the auditorium, it was completely full of teachers, staff members, and parents. A woman came to greet me, and she said, "You are Ana?" "Yes," I responded. And she said, "I know, you are the only one different here" (97% of the school staff and children were African Americans). "Come with me." She asked me to sit on the stage with two more people. During the time of the event, she introduced me as an expert on the Strategic Teaching and Reading Project (STRP), about which I knew nothing! The meeting lasted one hour. Those 60 min became the longest in my history of my professional life. I was the expert on something that I did not know! What if someone from the audience asked me a question? God help me! Once the meeting ended, I said to her. "Dr. Riggs, I do not know anything about STRP. Why did you call me an expert?" She looked at me, took the STRP manual and said, "I know that you can learn this from inside out." And, she left.

She was right. I went home, studied the entire manual, became knowledgeable on the content and strategies, and worked very closely with her all over the United States, different school districts, and diverse types of schools. When delivering a workshop, she, was an extraordinary and exemplary trainer, from whom I learned a lot. Dr. Riggs would say, "I am the trainer, Ana is the brain of the project." She transformed my professional life and in return changed my personal one! She took me under her wing. She sponsored me, and I will never forget her impact in my life.

The third anecdote brings me back to one of my dear professors, colleague, friend and brother, Dr. Charles Warfield, an African American who opened his home, with his dear wife, Dr. Martha Warfield, for me and my family. They both did not hesitate to embrace me as their sister. We had long conversations about my profession. I learned from Chuck and Martha what to make and take from bad moments in academia, especially those related to race and gender discrimination. They supported me in any decision I shared with them. Dr. Charles Warfield taught me to be comfortable with my language accent, with my mistakes. We talked about teaching methodologies for adults and the delivery of subject matter in the classroom. Chuck always thought that my risk-taking behavior, my unstoppable imagination and my fearless attitude were my strengths. He was one of the sponsors from whose advice, ideas, and feelings helped me shape who I am today.

Q. *In your own experience, what are the questions that the mentees should ask you?*

I do not have any preference, as all questions come from an inside worry. Some of the questions that I have been asked as a mentor include: "Do you think that I can do this? What do you see in me that makes you think I can do that? May I call you anytime? Could you write a letter of recommendation for me? What type of job do you think I can accept? Do you think I am ready for…?"

Also, "Could you share with me how you started in the academia? What are the obstacles that you have found and how did you overcome them?" I simply answer by urging them always to do more, to get involved. I have called people at midnight, just to ask if they have submitted important applications for which I know they were uncertain or hesitant. Through the years, people have appreciated these bold steps I sometime take to support and encourage my mentees or sponsored people.

Q. *Do you think single sessions, or mini-mentoring sessions are also effective?*

Single sessions are as important as mini-mentoring sessions. Mentoring can occur in any environment, for that reason, I prefer to mentor outside my office, at a coffee shop, for example, by phone, by email.

Q. *Do you think mentoring has to be in person, or are the on-line, digital mentoring solutions equally effective?*

Mentoring is a person-to-person/face-to-face relationship between the mentor and the protégé. However, in this era of technological modes of communication, I have mentored young professionals by using hangouts, zoom, skype, email. I cannot measure the effectiveness of my mentoring through those devices, but it seems to work as effectively as a face-to-face interaction.

Q. *Some people have had a special influence in our careers. Mentors and most particularly sponsors are some of those influential people. Please tell us about your own experience as mentee. Who mentored you in a memorable way? Who sponsored you?*

I would like to mention three. Firstly, Dr. Gene Thompson believed that my professional and personality traits (female, minority, Latina, doctoral degree) were an asset to any higher education institution in US. He was right to the point! The second is Dr. Ernestine Riggs, who believed that I was able to learn fast and trusted my ability not only to communicate with a different audience, but also to handle difficult situations. She trusted me. The third is Dr. Charles Warfield, who trusted my personal decisions. He admired my risk-taking attitude, my "no-obstacles on my way" behavior. He used to tell me "the sky is your limit, Serafin!" (Serafin was my ex-husband's last name during my doctoral program in Michigan).

Q. *Is gender or age important in mentoring and in sponsoring, females mentoring females, or more-experienced professionals mentoring younger people?*

Yes, gender and age are both important. I believe that those demographic factors are culture related. In the Latino community, we trust an aged person because we respect the experiences given by the course of the years lived. In terms of gender, Latinas trust Latinas, professional and successful ones.

If the mentor is a Latina woman, a successful one, the gender relationship becomes an extremely important connector between both, mentor and mentee. It happens because there is a social deficit in the US system in which successful Latina professionals are rare.

Opening a path to others, taking them under your wing, identifying others' potential are characteristics of sponsoring. Race and identity are two constructs important in my mentoring process as well. Being sponsored by African American and White professionals opened my mind and reframed my cultural misconceptions. They were older, culturally different, and experienced academicians in higher education and educational leadership. They had experiences that were unknown by me. Their level of academic expertise and exposure to life experiences influenced my life forever.

Yes, more-experienced professionals should mentor younger professionals. Unfortunately, today young people do not see the importance of that relationship. We, older professionals, are thought to have obsolete life knowledge. There is a saying, in the Hispanic culture, "*Más sabe el diablo por viejo, que por diablo,*" which relates to the value of professional expertise and personal experiences of older people when decisions are to be made. Yet, I look for that type of mentoring myself. There is value-added from age regardless of profession.

Online and virtual systems have created impersonal communications that are in-between a mentor-protégé relationship. I am not against technology usage, but I believe that technology cannot be the way for communicating feelings, ideas, excitement, etc. which are implicit in the two-way communication established between mentor/protégé.

Q. *And what about humbleness?*

What about being humble? What an interesting question! Some people don't think they need to be mentored. Some people come with the attitude that they know it all—not a person that you want to mentor. You need to be humble—to accept that the other person has experience.

And culture. Culture is also important.

Mentoring is culture-based. Latinas relate very well with Latinas. I reach out to more Latinas, because they see me as someone with some level of success. They can speak to me even if they have trouble with their English skills. They feel a level of comfort in communicating.

In the Latino communities, we trust an aged person, because we respect experience. If the mentee is a Latina, the relationship becomes stronger with a Latina mentor. Latina mentors or role models are rare and hard to find. Their absence is a deficit in the US higher education system.

I was a sponsored by two African American and one white professional colleagues. They opened my mind. They were culturally different. I think this was key for my own success.

Q. *What do you wish you would have done differently with regard to mentoring or sponsoring, as a mentor or as a mentee, as a sponsor or a sponsored professional?*

In relation to mentoring, I wish that I had been able to know how many mentees I had influenced for good. Many times, I received an email, a phone call, a person-to-person encounter telling me how I helped him/her or how my personal advice was followed and how success happened after taking my time to mentor someone. I cannot measure the impact of the transformation that I have caused on people. I wish my mentoring skills could be measured, but more importantly, I'd like to know how many times my mentoring has helped someone become a star!

I have taken under my wing, professionally many more Latina females than Latino males. I have spoken on behalf of many Latinas searching for employment or high positions. I have also fought for inclusion and equity for Latino males and advocated for Latino leadership parity at different levels of the society. I will continue doing so until I see that equity, equality and inclusion are respected in a society in which Latinx still face discrimination.

I create and look for opportunities for young professionals who I believe I should help succeed. Academic grades are only an isolated factor for me to use when mentoring or sponsoring someone. There are other skills, abilities, social and cultural knowledge that are more relevant when I mentor and/or sponsor a young professional.

Q. *Did you enjoy any particular mentoring/sponsoring experience? Why? Tell us more about those joyful occasions!*

I truly enjoy it when a young emerging leader calls me his/her mentor! I feel blessed that one way or another I have influenced his/her life. I enjoy meeting with my not so young mentees. I have a group of friends, all of whom were my mentees at some point in life, and we meet and laugh about how I would call them in the middle of the night and ask them to register for classes, because I was monitoring their registration and they were procrastinating! I would pick up the phone and without saying good evening say, "Registration for classes ends tonight."

"Do you think that the registration system will wait for you to remember to comply with your academic responsibility? Register immediately. You are so close to the finish line that I will make sure that you graduate whether you like it or not" and after a dry goodbye, I hang up the phone. They are now retiring school principals, school leaders, and are very grateful that whatever I did for them, it worked!

A Shared Selfie

- **Your favorite role model**: My mother.
- **One word to define your experience with mentoring**: Inspiration.
- **One word to define your experience with sponsoring:** Co-piloting. For me, sponsoring is a higher-level form of coaching—taking someone under your wing when you know that person will excel, like copiloting
- **Who would be an ideal mentor for you in this moment of your life?** Angela Merkel.
- **What would you ask this person?** What steps could I have taken in my life to build a political career like yours?
- **What question would you like this mentor to ask you?** What political skills do you have that make you think that you would be a relevant worldwide figure?

Afterthought

We were surprised to learn her mother is 90 years old, and that she was a singer. She recorded a CD with 12 songs for her 90th birthday. NBC went to Venezuela and interviewed her. Ana really admires her mother. She spent quite some time telling us about her. No wonder Ana is so energetic, she is daughter of a tornado!

Ana confessed she is a frustrated politician, and looks up to women in politics, envisioning them as role models. "*Political power is needed to do things you cannot do otherwise. She would like to be "a political figure for the well-being of others.*" She loves her working for others. She loves her life.

Dr. Ali Omar Al-Gheithy

"Sponsors are earned."

M. A. Capello and E. Sprunt, *Mentoring and Sponsoring*,
https://doi.org/10.1007/978-3-030-59433-6_22

A Glimpse

The Middle East region has produced oil and gas for more the 80 years, transforming what were small and underdeveloped countries into major players in the world economy. Oman is one of the six countries of the Gulf Cooperation Council (GCC). The others are Saudi Arabia, Kuwait, the United Arab Emirates, Qatar, and Bahrain. At the southeastern tip of the Arabian Peninsula, Oman has always held a strategically important geographical position as a trading hub, and now is the 25th ranked country in terms of oil and gas reserves. It is no wonder Oman has a long history of developing its human talent for its national oil and gas industry

Dr. Ali Al-Gheithy has spent his entire career in oil and gas, reaching very high positions in Petroleum Development of Oman (PDO), the national oil company of Oman, and now in Shell, as the Heavy Oil Contract General Manager for Shell in Kuwait. A promoter of best practices in hydrocarbon development, oil and gas production effectiveness, and organizational capability, Dr. Al-Gheithy is renowned in the Arabic Gulf.

Dr. Al-Gheithy is a leader with outstanding technical and inter-personal people skills whose insights into mentoring and sponsoring provide value to those he encounters along his path at increasingly higher levels each time, in a progression that seems unstoppable.

- General Manager of the Enhanced Technical Services Agreement KOC-Shell, for Heavy Oil Developments in Kuwait.
- Director of the Petroleum Engineering function at Petroleum Development of Oman, PDO, in charge of 2,000 people, he managed the company's oil and gas resources.
- Handled the planning and execution phases of the PDO Study Center for Hydrocarbon Maturation of Oil and Gas Assets.
- Cluster leader of the Lekhwair assets in PDO.
- Team Leader for development and WRFM of the Fahud/Lekhwair assets in PDO.
- Leader of a large cross-discipline team for Nigeria's giant field Forcados-Yokri.
- SPE Oman Section President, SPE Certified Petroleum engineer.
- Ph.D. in Petroleum Engineering, Imperial College, UK; M.Sc. Petroleum Engineering, Imperial College, UK; BSc in petroleum engineering from the University of Tulsa, Oklahoma, USA.

A Personal Snapshot

Maria Angela had the opportunity to meet Dr. Ali Al-Gheithy in PDO, Muscat, when he was the Director of Petroleum Engineering function responsible for hydrocarbon maturation, well and reservoir management, technology and tool as well as staff development including training-related purposes. Now, they are colleagues in Kuwait Oil Company. Eve had the pleasure of meeting Dr. Al-Gheithy at the first Society of Petroleum Engineers' International Petroleum Technical Conference, when she was 2006 SPE President and Dr. Al-Gheithy was the SPE Oman Section ex-President.

The Interview

Q. Please tell us about your own mentoring and sponsoring experiences.

I have been lucky to have been mentored for a long time before I started to mentor others as well.

From the beginning of my career, I have been mentored and coached. The best thing that leaders can do to invest in their young graduates is to share their pitfalls and experiences.

I have had several mentees over the years. After more than 30 years of partnership with Shell, PDO follows many of the Shell processes, including mentoring workflows. Shell staff in expat assignments rotate every four years, so I changed mentors every four years or so. Those mentors helped me develop my career.

In Oman, we had something called Mentoring Circles, which were initiated by Omani and expat females working in PDO. They were called "Hawa" and designed to mentor young graduates. In the early times of my career, PDO had only a few female petroleum engineers and now they form about 30% of the professionals, but more recently, we have had almost 50% female new hires for the last four years. A strong pipeline of female talent in the company will create in the future a more gender-balanced leadership.

The women wanted to be mentored by different directors in PDO, because there were not many women in senior positions. I was engaged in specific programs to mentor the Omani female talent, because the company was openly and actively pursuing gender diversity and inclusion. My mentoring sessions vary from one-to-one formats to a more common eight-to-one format, in which I mentor eight mentees. There was a dual-benefit: I would support them on what issues they wanted, and they would have the opportunity to know, to share experiences, and most importantly, to support each other.

I considered my key responsibility to be developing young people. The best way to do it is through mentoring.

To add to that there is an element of coaching. Which is important, especially for much younger talent. You coach on technical skills, but you also coach on strong skills (which were previously called soft skills). Topics include: how to deal with a difficult post, how to handle negotiations, and so on and so forth. As the mentee's career progresses, coaching on skills is left behind and career mentoring takes precedence.

Q. When you changed from technical management to a more strategic upper management function, were you mentored?

I received mentoring for that transitional period, and it was done by my line-management through a walk-about with immediate feedback. I had a mixture of Omani and expat mentors.

The most effective element in the mentoring I received was that mostly it was provided by someone outside my line. It is easier to speak-up to someone outside your line (chain of command). You do not have the pressure of being evaluated by the mentor, so there is more freedom to speak with someone who is not your supervisor. I consider it essential that the mentor is someone from outside your line or even outside your company.

There is an element important in mentoring, especially for the transition from technical to managerial roles. Willingness to pursue self-improvement and to seek your own opportunities is essential. Your opportunities will also depend on your track record, because people assess you on your merits when considering you for further opportunities.

Always aim to have a mentor who will guide you to do your job more effectively. It may be voluntary work, long-term planning or any other gap that needs to be closed for you to grow your professionalism and advance your career.

Let me provide one example. I don't speak up a lot. I speak very little. Through mentoring, I was advised to put my opinions forward more directly in meetings and to share my thoughts more freely in big events, when I feel I could add value. I was told that I should share my thoughts, rather than keeping silent and regretting later that I should have spoken up.

After this feedback, I put a lot of effort into that and it has paid off. It was achieved through continuous feedback: My mentors would tell me how I was—or not—improving, in what was a loop consisting of feedback-correction-action. Continuous learning with a feedback loop is the best approach.

You can also gain that kind of improvement using the 360° feedback approach, receiving the vision of supervisors, peers and supervised liaisons. In this way, you can become aware of your blind spots and try to learn from them.

Q. *Now, tell us about sponsoring. Have you been sponsored?*

I have been sponsored for sure. It is very useful to have a sponsor. For me, a sponsor is something you earn. You don't just get a sponsor. You earn the sponsor's trust based on your own merits, and your track record of leadership.

You may or may not be aware that someone is serving as your sponsor and has taken you under his wing. He or she will speak for you and open doors for you that may not have been visible or known to you. Sponsors will speak on behalf of you as a candidate, with statements like, "I have seen this guy perform. I have seen what he has delivered. Perhaps if we move him to Human Resources (HR), he will have a fuller, more rounded view of the company. We will need him or her to gain that experience to further advance." Those conversations and discussions do not happen in front of you.

Q. *Do you have an example of how you were sponsored?*

My sponsor shared with me his own experiences. He was a young Omani posted to Nigeria in the late 1970's and jumped two levels into management, because someone believed in him and wanted to stretch and test him. When he was telling this story to me, he said he wanted to do the same with me, to stretch me and then in turn, that that I should do the same for others later.

A good sponsor needs to support his sponsored individuals to succeed and not just to throw them into the deep end. It is key to stretch and support, when there is the gut feeling that "this person is different." You detect possible gaps in this excellent candidate and identify what needs to be fortified, what needs to be stretched. A sponsor is taking a risk when he says, "I'll put my neck out for him." or "Let's give him or her a chance," Those words can convince others in the decision loops to select this sponsored individual.

Usually we use the more routine way of developing people. Maybe new formats and specific sponsoring are needed to enhance the promotion of female talent that we have never had before. If someone says, "Let's give her a chance" and supports her, he may be able to convince the others that she is worth a try.

A leader will sponsor a person, knowing that promoting that specific person will most probably upset a lot of other employees, who are expecting to be considered and even selected for the role. A sponsor speaks about a candidate, who is not completely ready.

It requires you to have that special intuition, that gut-feeling, "This guy's talented. This fellow is different with something we need to stretch, to test. I will put my name out for him or her."

***Q.** What is the difference between sponsoring people when you are a technical manager versus when you become a deputy CEO, or a senior manager?*

It takes a lot of courage to be a sponsor, because there will be resistance from others, and they will question if they should let you take the risk or not.

One level is where you sponsor someone within your own team. This is the general case in technical sponsoring. You have more authority and the sponsoring process is contained within your own team. People in your team will also respect your opinion and they know everyone in the team, so, it is generally easy to reach consensus. Maybe, I should say it is easier to get consensus.

A broader and more complex sponsoring process occurs when sponsoring someone in leadership. The sponsor is putting his reputation on the line and not everyone involved knows every candidate. In these cases, you must present the case and support your reasons for advocating that person. If the candidate has a successful track record, that helps. You generally will gain the support of other leaders. Others will accept taking the chance not because they know the candidate, but because they believe in you. This is why it takes courage.

You put your neck on the line, when you advocate for someone. It is your prestige that is playing a big role.

***Q.** Do you want to share with us an example of this kind of situations of sponsoring at the top level?*

There are cases in which you use the more routine way to develop people. But when I was in PDO, there was a moment in time when we realized that we had female talent, but never had a female team leader. The momentum grew to have one and then discussions occurred around the need to advocate for our female talent. It was not done before. There were concerns expressed by teams accepting this leadership, but several directors took the lead, "Guys, let us give her a chance."

Q. *Have you done this?*

Yes, in Oman we had put forward a much younger female candidate for the post. People thought we were leap frogging, jumping her ahead two or three grades, but it worked well and female role models were finally there in leadership positions and inspiring younger females in the organization.

Another example I would like to share is my own case. When I came to KOC, I had to move out from PDO in February. The first thing my boss asked was, "Who will take your job?" I then had to convince my boss there were several people who could do my job. I convinced him that there were candidates who could do my job better than me!

I was asked to supply a list of three names of people who could do my job. That list went to the broader, 14-person executive committee. We discussed the top talent. Even if I sponsored someone, the candidates were to be discussed by the committee. So, this is an example of several sponsors advocating for your candidate. If someone has an outside sponsor, it gives more weight to the proposed candidate, and if that external, or second sponsor is senior, it gives it even more weight. Multiple sponsors are fundamental in this situation.

More sponsors are better. It serves you better. One sponsor may see some aspects. Several open many more doors.

Top talent in Shell is discussed by the leadership. My role was to present the petroleum engineering talent.

Q. *Have you noticed major differences between National Oil Companies (NOCs) and International Oil Companies (IOCs)?*

In PDO we follow Shell processes. So, I do not detect differences. I do not know the internal processes in KOC yet, so those may be different.

Q. *In your own experience, what are the questions that the mentees should ask you?*

Most of the questions I have been asked are pertinent to long-term career advice, including the usual dilemma as to whether to stay technical or broaden into other functions. That question occurs frequently.

Also, I receive many questions on work-life balance and how they can work effectively but have personal time. In the Middle East, many people, especially females will graduate, get married, have children and have a difficult time achieving acceptable work-life balance. I am specifically referring to the younger talent.

The questions also gravitate around how to deal with difficult locations and difficult roles at work. In my view, success in a career relates mostly with balancing long-term career goals and short-term goals.

*(**Comment**) When Maria Angela asked for his advice recently, Dr. Ali asked her, "Why is it that what you did in the past in a similar situation is not working now?" Maria Angela pointed out that when she asked a question, he replied with questions and asked, "Does that methodology of mentoring through asking questions come naturally to you?"*

In mentoring it is best to allow the mentee to "pull" because that is an opportunity for the mentee to ask questions. As a mentor, you must "hold back". If you ask questions rather than providing answers, the mentoring workflow is a winning situation, because you are really developing people. You are making them think.

They need to find their own way. When a mentee does not pull enough, you can offer options and guide, but you cannot steer.

On the other hand, coaching is more what you can drive for them. In mentoring, they must come to you and pull and ask for your support. When they identify for themselves what to do, it is more effective and you have reached your goal.

Q. *Do you think that single sessions, or mini-mentoring sessions are also effective?*

It is a journey, not a one session. It requires a plan and it must be a continuous journey, not a once off.

Q. *Do you think mentoring has to be in person, or are the on-line, digital mentoring solutions equally effective?*

Thus far, all the sessions I have experienced have been face-to-face, but we cannot ignore videoconferencing solutions, which may be equally effective if you know the mentee.

Q. *Some people have had a special influence in our careers. Mentors and most particularly sponsors are some of those influential people. Please tell us about your own* experience as a mentee or a sponsored person.

My managing director sponsored me to take a senior position in PDO by accepting that my leadership style was effective with my team, even though he considered my leadership style to be quiet and a bit passive. He thought, "This guy will bring the team together and not rock the boat too hard." He pushed me through to a very senior role for me at that time. It was an extraordinary opportunity for me and an extraordinary sponsorship as well.

Q. Is gender or age important in mentoring and sponsoring?

I still feel that a traditional sponsor is someone senior with an authority role within your own organization rather than someone outside your organization.

In terms of gender and age, more senior is more worthwhile than someone younger. Traditional sponsors for me are someone more senior.

I would say that the best sponsor is someone who speaks on your behalf without you knowing.

In the Middle East in particular, gender sponsoring is quite a problem. There are now many females in the workforce, but none of the current leaders are females. Look at KOC, there is not even one female deputy CEO, and out of 62 managers, there are only two or three female managers. There is a lot of capable talent coming up. Sponsoring females will really help. We almost need to be biased to be sponsoring female talent.

In Oman, where I know the situation very well, the mentoring cycle is driven by helping women and letting the diversity come through. Developing women is part of the KPI of management. The percentage of women in leading positions is a KPI. We are almost biased in promoting women and gender diversity.

A Shared Selfie

- **Your favorite role model**: Nelson Mandela
- **One word to define your experience with mentoring**: Transparency.
- **One word to define your experience with sponsoring**: Branding.
- **Who would be an ideal mentor for you in this moment of your life** The leader of Dubai, an extraordinary visionary, Mohammed bin Rashid Al Maktoum. I believe he is truly fearless.
- **What would you have asked this person?** I would asked him, "How would you navigate the difficulties of encouraging production of heavy oil around the world?"

Afterthoughts

When we finalized the interview, Dr. Ali spent some time explaining to us why he admired Nelson Mandela.

"It was very impactful to reflect how he used sports to unite his country. He took the time to learn the name of each individual player of the national rugby team and when he went to the stadium, he greeted them one by one, by name, thanking each one. The team united the country when they won the world cup of rugby. The sport united the country. … He was able to promote patriotism for the country irrespective of color of the skin. He was very clever."

He particularly liked this quote from Mandela, "I was jailed for 27 years, and I can still talk to these people, (so) why can't you? I forgive them."

Nelson Mandela frequently spoke of forgiveness. Here is one we liked from Nelson Mandela's *1990 Christmas Message,* "We must strive to be moved by a generosity of spirit that will enable us to outgrow the hatred and conflicts of the past."

Andrew A. Young

"People in top roles also need mentoring."

M. A. Capello and E. Sprunt, *Mentoring and Sponsoring*,
https://doi.org/10.1007/978-3-030-59433-6_23

A Glimpse

Like a significant number of petroleum engineers, Andrew A. Young's training was in chemical engineering. He began his career path in Australia and rose to global prominence as a leader. With more than eleven board memberships over his career and having been the driving force behind the public offerings or public listings of more than ten companies in the energy and finance consulting area, Andrew is an expert, who is constantly called in to provide his insights in an advisory capacity including as a legal and expert witness.

His long trajectory includes having served as the first President of SPE from the southern hemisphere. Among his accomplishments as SPE President were global expansion of SPE activities and the launch of the Middle East office of SPE in Dubai. Andrew emphasized that SPE should be a truly international group, not just a professional society with its core in the United States and a few overseas outposts.

- Board appointments at Gaffney Cline & Associates, Anzon Australia, Anzon Energy, DigitalCore, New Guinea Energy, Galilee Energy, Cue Energy, SPE, NSCA, Hoqool Production, Cliq Energy, Real Energy, APPEA, and Chalro Oil LLC.
- Led the Public listings of Anzon Australia Limited (ASX 2004); Anzon Energy Limited (AIM LSE 2005), CLIQ Energy (BURSA KL 2013).
- Managed business contracts and services in Australasia, USA, South East Asia, Nigeria, Middle East, Khazkhstan & Argentina.
- Renowned consultant on technical, operational and managerial issues to independent and national companies and government organisations, with a successful track record in Nigeria, Iran, Saudi Arabia, Bahrain, Kuwait, UAE, Iraq, Syria, USA, Indonesia, Malaysia, Thailand, Singapore, Hong Kong, New Zealand, Papua New Guinea, Timor Leste, Khazkhstan, Argentina.
- Founder of the Gaffney Cline & Associates practice in Australia.
- Honorary member of SPE.
- Instrumental in forming the first School of Petroleum Engineering in Australia at the University of NSW in 1985, and Chairman of the Advisory Board 1988-2011.
- Councillor with the Australian Petroleum Production & Exploration Association Limited (APPEA) 2005–2008.
- Bachelors of Chemical Engineering from University of Melbourne, MBA (Hons) from University of Rochester New York.

A Personal Snapshot

We were keen on interviewing Andrew, as one of the most effective sponsors of women at the SPE.

Eve had the privilege of meeting Andrew when they were both serving on the Board of the Society of Petroleum Engineers (SPE) in the early 1990's before everyone was talking about mentors and sponsors. At that time women were a very small minority in the petroleum industry and faced numerous barriers to full participation. Andrew was always a perfect gentleman and strong champion for women.

Through informal chats over the years, Eve has received a lot of important guidance and support from Andrew. He was instrumental in sponsoring Eve to become the 2006 President of SPE, which has had only five women presidents thus far.

Eve shared with Maria Angela the huge impact Andrew had on her life. One day in late 2003 or early 2004, Andrew called me up out of the blue and asked if I thought I could serve as SPE President. I told him, I thought I could, but would have to get permission. He replied there was a lot of competition, so I might not get selected and it would be easier to get company permission after if I was the person chosen."

The advancement of women in pioneering times takes courage, and we are proud to include Andrew's insight in our compilation about mentoring and sponsoring, as they provide a valuable historical perspective worth sharing.

The Interview

Q. *Please tell us about your own mentoring and sponsoring experiences.*

Mentoring is very important for me. I do not really differentiate between mentoring as a process initiated by someone approaching me with questions and what I experienced in my career. As I progressed through the ranks of an organization, I found that when I communicated with young people, there was always an element of mentoring.

Specifically, I have been lucky in that I have had variety, working for short periods for some companies and longer periods with others and in many countries. These companies have varied widely in terms of size and the focus of their business. This has given me the ability to compare and to contrast different experiences to gain a rounded perspective on work.

My first position was with Exxon for 8 years. In this early phase of my career, I was in a role of Junior Engineer, and I was nurtured by more senior people. Then, as I grew as a supervisor and as a manager, I actively mentored and supported the empowerment and growth of young people.

Q. *In terms of being mentored, is there any experience you found to be extremely valuable?*

I was very fortunate when I joined the workforce, as my first supervisor was a very dedicated mentor for me. He subsequently moved on but maintained contact and for 12 years I had the same mentor. He had a big influence impacting my advancement, professionalism and enjoyment of my job. His impact on me, while not a direct supervisor, helped a lot in receiving information and guidance about a desirable career path.

As a mentor myself, I have been particularly pleased to see my mentees advancing in their careers and after becoming high-rated performers, still seeking my advice and insights. There is enormous personal delight for me in seeing young engineers develop professionally.

Q. *How many people did you influence as a mentor?*

Some people may see mentoring as their interactions with younger people directly under their supervision, mixing mentoring with professional supervision. Accordingly, I have "mentored" many people in my career.

You must make sure you differentiate between internal and external mentors. Internal mentoring processes are those entirely within your company. Internal mentoring in large corporations and in small companies is an interesting topic to discuss, because technical and line managers benefit from working with employees who have received appropriate mentoring. It inherently shows in their employees' performance.

Q. *You mentioned external mentoring. What do you mean by that?*

Mentoring has not always been facilitated within organizations as it is now. Many companies now have mentoring as an assigned task.

In contrast, you can benefit by finding a mentor outside your organization. There are always opportunities for external mentoring. In the "ancient times" meeting at large industry conferences or professional societies' annual conferences were the main way to network and meet with people outside of your company.

I find after a career span of 45 years that technology should be acknowledged in this discussion. Now, we are in an age of mentoring. Previously, the process was time intensive. We started by emailing each other after initial meetings in person. As technology evolved mentoring extended to people who we encountered through e-conferences, webinars and video conferences. With these modern communication methods, mentoring has dramatically expanded and is orders of magnitude more common than when I started in the 1970s and in the 1980s.

The feasibility of 'unsolicited' mentoring and communication has also developed in the last 15 years, triggered by the introduction of LinkedIn, Facebook and other platforms. Today, you can reach out to people via social media, email and free-of-charge video calls. Because of my experience, I get unsolicited approaches from young professionals quite often. I would say that every month, two or more individuals reach out to me asking me if I can be their mentor. These are people whom I do not know!

SPE, the Society of Petroleum Engineers, was and still is a tremendous conduit. After I gained about ten years of experience, people reached out to me for advice, commentary, and suggestions.

Q. *At the beginning, mentoring is often combined with technical guidance, but for the purpose of our book, we are more focused on career mentoring. Tell us more about your views on this. How do you envision the evolution of mentoring along a career path of an individual?*

Early in my career, the individual to be mentored was usually a colleague at the organization where I worked. To start I would raise questions. I am not talking about purely technical questions.

Let me share my own experience. Back in the 1970s, it was uncommon for young engineers to work as an operator on drilling rigs or production platforms. I saw that as an opportunity to communicate with operational people and to gain operational experience that otherwise I could not have. The idea was completely left field. So, rather than raising a topic that might be career limiting, I discussed it with my mentor. I was lucky he was kind in his mentoring and also revolutionary.

Q. *Are you saying drillers were only working in office or just on a drilling rig?*

Yes, in my early days, it was almost a norm for a drilling or production engineer to be writing programs from behind a desk. We changed that paradigm.

Q. *Now, tell us about sponsoring. Have you been sponsored to advance in your career? We are referring to that process with which a person in a role that enables decision, advocates for certain individuals to be promoted. A case of specific sponsoring for example, started with women in the 1970s or 1980s. The industry realized that women were not getting relevant roles in organizations and would* *require sponsors. People who specifically advocated for them to increase their visibility and to overcome unconscious biases of senior management or members of the Board of Directors, who were not accustomed to seeing or promoting women into leading roles.*

Yes, I have been sponsored and have been a sponsor in this sense.

After I left Exxon, I joined an Australian mid-size company. This company had a corporate internal lawyer. Very good lawyers are often women with their attention to details and capability for multitasking. I thought my new company was a little biased against women and we were limiting the ability of the professional women to progress. The company even excluded capable female leaders if they had children. I became quite outspoken about this. I asked" Why can't we have female incumbent leaders?" I think I triggered reflection and action.

We implemented a system which allowed women to opt for part-time responsibilities during their pregnancy and early years of maternity. The arrangement of working part time was revolutionary at that time. I assured my colleagues at the management posts, that if they got the job done, we should enable part-time schemes of work. In 1986, it became the norm in that company. Since then in every company I have run or worked with, I have made a point to offer part-time work as an option. In fact, I have often formalized two 50% part time women employees making up 1 full time equivalent. This is generally even better than one full time person in the role, particularly corporate law.

Q. *Would you let a guy do it too?*

Yes, of course.

Q. *In your own work storyline, do you remember when you were sponsored?*

I want to turn the clock back to 1990. I, as General Manager of Operations and my colleague, the Finance Manager of the company, were both asked by the CEO of the company whether we would be interested in pursuing a Master of Business Administration, an MBA, sponsored by the company. The company would give us the time and cover the tuition and living expenses. Imagine that! Two General Managers in a full day class per week! We said yes, as we felt honored. There were no strings attached. Our life was busy however!

Q. *Who was your super sponsor?*

The CEO of the company. He was the one behind the idea and convinced this was an opportunity and clear path to develop the senior management of the company.

Q. *In your own experience, what are the questions that the mentees should ask you?*

There are 3 components to this question.

First, the educational element. I consider proper questions to be those related to the educational profile, for example, "What additional education will I require after my professional training at the university? Would you recommend that I do some sort of finance degree? Should I pursue an MBA? What sort of education pathway should I pursue?" They should ask about Internal professional development as well as external development at their own cost and time. I don't have a cookie cutter for that sort of question. It varies with the individual, but education is key.

Secondly, discussion of jobs and taking responsibility is important. I am a very firm believer of seeking roles within the company which have a differentiator and can be completed in a practical timeframe, i.e. within a year. Anything more than a year is not necessarily achievable. Undertaking high profile tasks, which are particularly relevant is important if you work in a large company. That is where people seek a mentor. Generally, they are ambitious. Taking a job that has responsibility and therefore an opportunity to make an impression is very important to somehow to stand out from the crowd.

The third element is to make yourself visible by writing papers. That helps to develop pride and reputation. It helps establish your technical strengths. The papers can be internal or external and may require an internal or external mentor.

Q. *Do you think single sessions, or mini-mentoring sessions are also effective?*

Yes, mini-mentoring sessions are effective, but only for specific questions. A plethora of topics cannot be covered in a single or mini-mentoring session

Q. *Do you think mentoring has to be in person or are on-line, digital mentoring solutions equally effective?*

Obviously, you want to have a personal relationship first. When you know what the person is seeking, then and only then can you move to e-mentoring. That type of mentoring is particularly relevant today and FaceTime or Skype works very well.

Q. *Are you currently mentoring someone on-line?*

Yes, a number of people. I just mentored someone in Sydney on-line in an e-mentoring session.

Q. *What roles do you think professional societies should play for e-mentoring programs?*

The role of professional societies is to draw members' attention to the diversity and plethora of systems and processes that are available and to provide a platform or medium so that members may engage in e-mentoring at minimum cost.

I don't think it goes beyond that. The cost metaphor, "You can lead a horse to water, but you cannot make it drink."

The role of any professional society is to make use of the experience in the membership to improve techniques to help develop young people.

There is an interesting situation which I would like to share. The irony of my story resides in the very first question you asked about a mentor who had a big influence on my career. Well, in Exxon, what I had was almost a gift from heaven. I am a chemical engineer. I did not pursue petroleum engineering, because the discipline did not exist in Australia at that time. However, chemical engineering is a building block for petroleum engineering.

I had no knowledge of the oil and gas industry, other than understanding chemical processes. During my final years of university there were campus interviews in which a young company representative came to interview people. The person who interviewed me was the same individual who later became my mentor! He had an extraordinary magnetic influence on me to pursue oil and gas as opposed to any other industry. I owe this person enormous gratitude not only for influencing me to go into petroleum engineering, but also for subsequently helping to develop me as a person and as a professional.

Q. *Is gender or age important in mentoring?*

I would like to answer with an emphatic and categorical, no.

Experience may be proportional to age, and yes, experienced people should mentor less-experienced ones. In terms of sponsoring, I would apply the same logic.

Q. *Men sponsoring women may be perceived to be more than a merely professional liaison, do you have any feedback on this?*

If there was any sort of seduction involved from either side that would imply some sort of favoritism. Personally, I have never experienced that kind of a problem. The irony is that when there is hyper-sensitivity, unfortunately that is read as discrimination.

So much of the world today is highlighting elements of inappropriate male-female relationships, that high sensitivity is present. Some organizations now require a third party in the room when a male is talking to a female. I think that is hyper-sensibility which can be limiting for a women's career. In Latin America there is more flexibility in relation to the liaisons among men and women. I have seen a different approach in Latin America, but I don't know about that in the Middle East.

(Comment) *Andrew asked Maria Angela about sensitivity due to this matter in the Middle East.*

Maria Angela responded, *"The situation is very dependent on every cultural context. It is different in the Middle East. I see women fight for their rights at work. The difference is that generally, they have no sponsors, because it is awkward for both sides of the deal - for a man to openly support a woman and for her to receive a strong endorsement from a man. There are no connotations of any sexual-related issues of which I am aware. I feel very safe, and perceive my female colleagues feel very safe in this sense."*

Q. *What do you wish you would have done differently in reference to mentoring or sponsoring, as a mentor or as a mentee?*

People did not talk about formal mentors or mentees at the beginning of my career. It is a new concept in many ways. Therefore, we are talking about it right now.

If I could live again, I would have consciously sought an external mentor.

I didn't have an external mentor until many years after I started. I mentioned I had one internal mentor for my first twelve years in a company I had another mentor for twelve years more and then, another for about another six years as I moved from company to company.

When I became Managing Director there was a hiatus of about six years without a mentor. I can share that it was a lonely space in the mid-1990s. When I was and felt alone. I was moving from pure technical to pure board positions and I found it difficult. There was no one to whom I could talk about it. It was a lonely situation. I wished I could liaise with a very mature professional, who could empathize with me as a mentor. Some people think you have finished, when you have arrived at the top, and you no longer need mentoring, but I disagree. There is often a time when I want to have someone to whom you can talk. It can be very lonely. You need people whom you can trust.

As a board member, I do not look at my peer board members as mentors They are like colleagues, and you must be careful not to show your cards up your sleeve. A short answer. The tricky part of life is when you get up in the higher levels of management and you are expected to know everything, you cannot fail. You must inspire trust.

In the last few years, I have found mentors in highly intelligent people whom I respect. One is a lawyer, who has become a friend. I met him because he was the father of one of my children's best friends. We have become very close and although we are family friends, we also mentor each other. He is my sounding board. That is a new form of mentoring for me, same age, no conflict of interests. The second person is my wife.

It is healthy to have someone with whom to kick around ideas.

A Shared Selfie

- **Your favorite role model**: I have three.
  - My Mother-in-law, who is now 94 years old, and seems to always identify the positive side of things. In more than 50 years, I have never heard her say anything negative about anyone. She is a very positive person like no one else I know.
  - Peter Gaffney of Gaffney Cline & Associates. He is the person I was speaking of being one of my mentors.
  - My father, who taught me the values in which I believe. I am not a religious person, but I follow high values/principles and try to convey those in my work and in my family.
- **One word to define your experience with mentoring**: Sharing.
- **One word to define your experience with sponsoring**: Facilitating.
- **Who you wanted to be your mentor/sponsor, but you never had the chance to ask?** My grandfather, who passed away when I was fourteen. I admired him. He was a very successful accountant, head of faculty at Melbourne University and advisor for the government during the WWII on managing of the finances of the country. He was an extraordinary person I would have loved to have my grandfather, Sir Alexander Fitzgerald, as my mentor.

Afterthought

Andrew's interview was an extraordinary one in many ways. Including the fact that when we sent him our draft, to review for any inaccuracies, he replied he had just came back from the hospital after three days in the Intensive Care Unit, and a total of 10 days, to recover from an accident.

While on his early morning bike ride he was wiped out by a car. The result was he had 5 broken ribs, 1 broken clavicle, 1 smashed vertebra, 2 very large hematomas of his torso, 1 very bruised and sore left hip replacement prosthesis, and he smoothed his email saying: "*I am typing this note to you personally and it is not an angel from heaven!!*" (sic.)

Some leaders are made with steel fibers, and an excellent example is Andrew.

Dr. Susan Jane Webb

"Learning to listen has been fundamental for my mentoring"

M. A. Capello and E. Sprunt, *Mentoring and Sponsoring*,
https://doi.org/10.1007/978-3-030-59433-6_24

A Glimpse

Sometimes, a person migrates to another country and finds in it a new home, which expands their horizons and positions them to benefit others. Dr. Susan Webb, an American who has made of South Africa her home, strives to enable students to become what they aspire to be. In sharing her insightful knowledge of geophysics, Dr. Webb has greatly advanced the study of Geophysics at the University of the University of the Witwatersrand. The support, praise and even love of her students is clearly apparent and many of them told us proudly they were positively motivated by Dr. Webb's guidance.

Dr. Webb has authored more than 150 technical articles and peer-reviewed papers in many of her discipline's most important journals. She has given presentations at geophysical conferences around the world and created a viral YouTube video, "Baby elephant calf vs birds," that has had more than 28 million views and has raised money for my university Geoscience Alumni Scholarship. She is a natural mentor and sponsor of many young professionals who speak highly of her.

- Ph.D. Geophysics, University of the Witwatersrand (South Africa); M.Sc. Geophysics, Memorial University of Newfoundland (Canada) and B.Sc. Geophysics, SUNY Binghamton (NY) (USA).
- Associate Professor, School of Geosciences, University of the Witwatersrand, Johannesburg, South Africa; and Director: Africa Array/Wits International Geophysics Field School.
- Multiple appointments as a Senior Lecturer in South Africa and Canada for universities and Mining organizations. Researcher at MIT and Carnegie Institution of Washington.
- Executive roles as member in the Board of Directors of the AGU (American Geophysical Union), SEG (Society of Exploration Geophysicists), founding member of Geoscientists Without Borders (GWB).
- Advisor, Supervisor and Examiner of more than 150 thesis and dissertations of Master and Ph.D. level.
- She has been the recipient of many awards including, Meritorious Geophysical Service SAGA, SEG Outstanding Educator Award, SEG Honorary Lecturer for Africa and the Middle East.

A Personal Snapshot

Dr. Webb is very well known in the worldwide geophysical community, even though she has been a resident of Johannesburg since 1990. She has worked to enhance respect for minorities in the international geosciences communities and championed beyond-the-main-stream geophysical technologies and applications, including gravity and magnetics for Mining as well as for oil and gas.

The Interview

It was easy to organize the interview with Susan, because she is a friend and a colleague, and is alive in many social media channels. Her experiences provide a fascinating perspective on mentoring.

Q. Please tell us about your own mentoring experiences.

Most interesting for me, has been my immersion in different cultures. As an American woman with a many male students, I was in an unusual situation and I immediately discovered I had to learn to listen. I dealt with many first-generation university students, who were also the first high school students in their families. They learned the path to higher education by walking it. Their issues are different from mine and they do not know what mentoring means. As the first in their families to reach certain academic levels, they raise questions that are not all mentoring issues. Often, they lack the social skills necessary to succeed in multicultural or academic communities.

A lot of learning is achieved by listening. I help them learn what they want to do, because they may not be assertive enough to take on opportunities.

Q. What do you mean when you say there is need to establish boundaries?

I had to explain what mentoring is and establish boundaries and roles. Many students I have do not understand the difference between mentoring and supervising.

For me, it is about understanding how people develop themselves, and empowering them to do that.

People come to me with all kinds of problems, but I am not a psychologist. I have found myself in situations where people asked me for permission to do things which we consider a personal, non-professional decision such as marriage, travel and vacations. Often, they mistake mentoring for supervision or parenting, which are different roles. Women, in particular, would come to me with problems that I do not consider mentoring.

Q. *Give us some examples.*

They would ask questions such as, "How do you exchange this purchase?" That is certainly not within my role as an academic mentor. They do not have the social or networking skills they really need.

It is important to figure out where you can add the most value. The goal is not to tell them what to do, but to help them figure out what they can do. My goal is to point them towards opportunities, some for which they may not be ready.

I cannot directly encourage them, because you cannot provide courage. They must find and build courage in themselves. It is for them to determine what is their attainable and preferred choice. It is about helping them to reflect on their skills and choices.

Q. *In your own experience, what are the questions that the mentees should ask you? Or in other words, what questions do you want to be asked in a mentoring session?*

I think the questions should be "what opportunities are out there and how do I take advantage of them?" Mentorship is about guidance, not someone telling you what to do. It is up to you to decide.

A valuable question could be "Do you know anything about … (an organization, a role, a country, an industrial sector)? In other words, I like questions that ask me to specify details about opportunities or to share my own experience and insight on elements related to that opportunity, so that the student has a more complete understanding for her decision.

(**Comment**) *Eve shared that early in her career, she never had any formal mentoring. She said, "I feel close to your students who do not understand the mentoring process at all. It only after years of working that I realized what mentorship was or the importance of sponsorship. I only had yearly reviews of performance from my supervisor. I realized there was a need for explaining the differences between mentoring and sponsoring, because it is something that is not frequently talked about."*

Q. *Did you have a mentor or sponsor when you were a young professional or a student? How did you learn to be a good mentor?*

I did not have any mentors as a student, only very bad experiences that I do not want to share at this moment, but I did have some people from whom I received good advice.

Q. *Did you have the concept of mentor/sponsors when you were a student?*

I did not realize that until mid-career. I found people who helped me and gave me advice about staying in the academic world. Also, I experienced a terrible bullying situation—a bad dream I may someday talk about, but not now.

The concept of mentorship and sponsorship has come in the forefront in career development, previously it was hidden and unknown. It is a much formal and recognized process than 40 years ago.

Q. *Do you consider yourself a good mentor?*

It has been a learning process for me. I had to learn how to be a mentor. What did it mean and how could I add value? It is different from being a supervisor, even if those roles at times get blurred. It is all about opportunities and putting them in front of people.

Q. *Have you mentored people outside of your organization? That is a pure mentoring situation, just advice, not influenced by rank.*

Yes, I do get many young professionals reaching out to me. Unfortunately, I have had to learn to say "no." However, even trivial meetings over a cup of coffee may be transformed into a mentoring session. Then, people come back, to you.

Q. *Are mini-sessions or one-time mentoring sessions effective?*

Single sessions may be valuable, but I think multiple mentoring sessions is how mentoring works best. Students find what they want and dig for it. They can tell if people have the information they are looking for or not. Here in South Africa, women, especially early in their careers, can be very determined, seeking specific connections and opportunities.

Q. *Have you tried mentoring on line?*

Digital can work, and I have had quite good sessions, but it is hard to ask difficult questions when it isn't face-to-face.

I have done a fair amount of mentoring through Facebook and LinkedIn, I always try to respond. After a while people let me know it was really effective, even if the contact was through mobile phones text messages. This happens if and only if people ask the proper questions. I figured this out in my process of learning what mentoring means. Once the mentees figure out what they want to do, or achieve, they are effective finding resources and asking questions. They do not lean on you too much. Once things are figured out, they move on. Perhaps, they will use a mentor for a short while, and then move to someone else. It is very dependent on the objectives at stake.

Q. *Have you had long-time, life mentors?*

I have had a few. I have also had a number of liaisons in which the relation evolves from supervisor to mentor, and then to colleagues and finally, true friendship.

In a small and specialized field such as mine, things tend to go in a circle. Situations evolve. For example, I started with a student in Honors College who has now become a colleague. Another person who I mentored as a student for her M. Sc. and Ph.D., now has a Post Doc assignment, and I have become a long-time mentor for her. I help her to envision what she wants and ask for opportunities.

Q. *Let us shift gears and talk about your capacity as sponsor. Have you been a sponsor for anyone?*

In academia, sponsoring is not so easy, because the processes are designed to be based on merits and numbers. But I have certainly spoken up for qualified individuals who are remarkable and way above their peers, and I advocate for them. I have done that quite a bit. Maybe that is like sponsoring. In academia, there is a lot of structure and paperwork, but I always make sure my recommendations are there for deserving individuals whom I care about.

Q. *And about your own experiences being mentored?*

It is kind of funny. Many mentoring sessions have been quite helpful for me. Even as far back as when I was an undergraduate, I remember being frustrated in geophysics. Then, I had a casual session with a friend in graduate school, who pointed out other career options in geophysics. I understood and connected with these opportunities in the field! It was one of those chance conversations that was a valuable mentoring moment. Shining!

Q. *Do you receive mentoring now? Is mentoring something that accompanies a person all life long?*

We have developed in our small group a culture of weekly meetings that is our substitute for one-to-one mentoring. We do this as a group.

Q. *Is age important for effective mentoring and sponsoring?*

I'm glad you asked that, because I think that age is somewhat interesting. I have had quite a bit of mentoring from someone younger than me.

As someone from a different culture, I may do things that may come across as offensive without realizing it. I cannot really ask my students, so I ask my younger colleagues about it. It is an interesting relationship. I am asking, for example, "*What am I doing that seems to you to be racist or sexist*?" These are difficult issues to discuss and they must be spoken about in a very respectful way. I recognize that my younger colleagues are more aware of the cultural concerns because I come from a different country and culture. They are very respectful of my scientific profile.

Some of these cultural conversations have been unforgettable. Mentoring by a younger person has provided insights on multiculturalism that are incredibly powerful and interesting.

Q. And about gender? Is gender important in mentoring and sponsoring?

It can be important, but it is important to get both perspectives and see both sides of the coin. A male mentor may help a female to better handle difficult conversations at work. I feel fortunate to live in a place where I can have these conversations and move forward and as colleagues. There are difficult issues related to career advancement, women's self-empowerment, sexual harassment, and different ways to mentor people about these difficult topics and challenges. People will find different mentors for different aspects, and I think guidance from men and women is distinctive and important for both men and women.

Q. What would you have done differently as a mentor?

I wish I understood what mentoring was many years ago. This is because I was not really mentored. At the beginning of my own experience as mentor, I think I did not listen well. I was advising, not mentoring. The goal of mentoring is to open opportunities and trigger reflections in the mentee.

Is hard for me to say, "No." There are times I wish I had just said "no," because I did not have the time for mentoring sessions and follow up. I had to learn to say "no" in ways people understand.

Q. And what would you have done differently as a mentee?

As a mentee, I would have looked for a variety of mentors. I now realize I was mistakenly looking for friendship instead of mentorship.

Q. Should there training in the university curriculum on mentoring?

No. I think we need to create opportunities outside the curriculum for the students to understand they must be CEOs of their own careers, seeking the appropriate resources. We must stop making everything needed to shape the professionals of the future part of the curriculum. It would become yet another course and the attitude of "*I have to go through this*" is dangerous. Mentoring is so important, they must think about it. You do not want mentoring to have a grade, to be part of a GPA. It should be part of how to manage expectations. Mentoring is not required, but I would love to see opportunities presented to every student and every student seeking mentors.

(*Susan Webb showed us during the video conference a coffee mug engraved with her name.*)

I am very proud of this mug because it is from a program run by students, called "Bridge the Gap." It is a peer–to–peer mentoring program. They gave me the mug with my name on it and it is precious for me. They run this program themselves. They are strict and do not want the professors involved. They come with questions. This program truly inspires me.

They ask for mentoring themselves, as they start to look for work. The interview process has changed, and even if a resume gets your foot in the door, an interview is much more key to get you the job, much more than it used to be. We don't talk about that enough. The whole process is grounded on emotional intelligence, which is not sufficiently understood. I consider corporate culture is border lining on racism, because it blocks people from joining. I have seen that happening with my students when they are fresh graduates seeking job opportunities. Sad.

Q. *Do you think how to be a mentor can be taught?*

I think you have to be careful. You cannot push everybody to have the same amount of time for mentoring. For example, in academia, we have 30% admin, 30% teaching, 30% research My point is that we all have different skills. There are people good for different types of mentoring. I may not be a good mentor for certain kinds of people. Is important to make sure people are expected to do mentoring, but we need a cautious mix-match approach.

Q. *Can you tell us your point of view about the difference between a role model and a mentor?*

Role models are incredibly important! We notice that on a daily basis here. As we grew our own world-class role models in South Africa, it made a huge difference, completely changing the conversations.

In the personal sphere, I must confess I had a wakeup call once, coming into work. One of the female scientists stopped me and said, "I want to be just like you!" I realized it was OK to be a role model. It is OK for people to look up to you. That is not what people necessarily want, but if you are experienced in your field, people will look up to you.

Q. You are the author of a viral video on YouTube. Your voice is heard by many on Twitter and Facebook. What are your thoughts about the "jealousy" in social media, with not all "liking" or "reposting" interesting materials or achievements of colleagues?

I see there is true fear in the use of social media. If you "like" the wrong post, you may get fired! I give my "likes" to lots of people in social media, and that does not bother me. What bothers me is why there aren't more mentoring opportunities in social media. Social media is a platform and good at connecting people. My colleague, Stephanie, has a good YouTube channel, with over 100 K followers. How do we promote learning and exchange of knowledge in an on-line environment? How may we effectively use it for mentoring? I would love to see global initiatives grounded on social media. For example, how to quantify the uncertainty all over the world, because every culture has a different perception.

A Shared Selfie

- **Your favorite role model**: I have several. Louise Pellerin, Maria Angela Capello, Carol Finn, Chris McEntee.
- **One word to define your experience with mentoring**: Opportunity.
- **One word to define your experience with sponsoring**: Support.
- **Who you wanted to be your mentor/sponsor, but you never had the chance to ask?** Margaret Layman. She was on the AGU Board of Directors with me.
- **Who would be an ideal mentor for you in this moment of your life?** I was very lucky when I was younger as I was in the Olympic horse-riding team, and there was a person that even now, is a mentor for me. I wish I could be closer to this person. I cherish his long-distance mentorship and long for more.

Denny Emerson continues to be an inspiration and mentor on social media…and if it would be feasible for me, I would reach out to him to be my mentor.

Afterthought

This particular interview carries a positive emotional connotation for our book, because it was the very first interview we did. It was conducted virtually, joining from Rome (Italy), Johannesburg (South Africa) and California (USA).

We always find Susan Webb to be a supportive collaborator. She is a role model in more than one way, and we were thrilled to include her views in our compilation.

David G. MacDonald

"Mentoring is about building a relationship."

M. A. Capello and E. Sprunt, *Mentoring and Sponsoring*,
https://doi.org/10.1007/978-3-030-59433-6_25

A Glimpse

One of the most important workflows in the energy industry is the assessment of resources. This grounds the forecasts for demand and sourcing of all the resources in the energy basket. In the oil and gas segment, this is a key step and one that positions companies and countries in the economic scenarios.

David MacDonald has dedicated a good portion of his career in the industry to the assessment of the international resources of BP. He is charged with ensuring the compliance and integrity of BP's reserves and resources. Not an easy task in a time of economic turmoil of many kinds.

- Leads a global team of engineers, geoscientists and finance professionals to ensure consistent and compliant estimation of BP's reserves and resources for business planning, resource progression and government reporting across all of BP's regions and functions.
- Fostered and led the development of the first multi-commodity classification system for extractive, renewable and anthropogenic resources with the goal of supporting the fulfilment of the UN Sustainable Development Goals.
- Champion for gender diversity across the extractive industries. Co-chair of the BP women's network for the International Centre for Business and Technology. Introduced policies across BP for recognition of the impact of women's health issues, particularly menopause, on career development and business success.
- Baccalaureate in Petroleum Engineering from University of Oklahoma.

A Personal Snapshot

David was known to us through his prestige, and some common liaisons.

During the interview, we noticed his passion for sailing, an element that grounds some of his views on mentoring.

David MacDonald has also been deeply involved in the diversity and inclusion programs of BP, specifically for gender-diversity, and because of his insights on the topic in one of the largest corporations of oil and gas in the world, we were keen on including his views in our compilation. An internationally recognized expert on the estimation of reserves and resources, and current BP Vice-President for the Segment Reserves in the Finance Function, he is based in the UK and accountable for the estimation, categorization and reporting of BP's oil and gas proved reserves.

The Interview

Q. Please tell us about your own mentoring and sponsoring experiences.

I am very engaged in gender issues. I am the co-leader of BP's Women's network. I think that once you become a senior level leader within a company, your value to the business is what you can give younger or less experienced employees. Doing technical work is of less importance. My focus is on looking for opportunities to help young people excel in their careers.

In relation to mentoring and sponsorship, I think there is a little bit of overlap. It is difficult to be a sponsor without being a mentor, but you can be a mentor without being a sponsor.

A mentor intervenes and may provide help in how young or inexperienced professionals can progress their careers. It is about giving good advice. As a mentor, there is an element of risk, because if that advice is not followed, it does not produce the desired results.

While there is some personal risk in mentoring, there is far more risk in sponsoring, because there is a level of accountability in sponsoring with regard to the success of the people you have recommended. If I advocate for someone, I need to have some level of assurance that this person can deliver as he or she progresses. I must see that the sponsored people are progressing well, because that has an impact on my own prestige. I take on some level of risk when I sponsor.

Is difficult to be a sponsor without knowing the skills of the person you are sponsoring, and without going through a mentoring process as well.

Mentoring is about trying to take into consideration not just my work and life experiences when I give people advice about how to navigate their careers, but also to learn how others navigate as their environment evolves.

For example, providing advice to women is challenging because the environment has changed so much. The advice of senior women may not be relevant anymore for the new generations. In BP, senior managers are invited to speak to younger ones, but those experiences may no longer be relevant.

At the early times of organizing activities for empowering women, the easy thing to do was for the women's network to have someone talk about their experiences, but in our company, we realized it was not really irrelevant to women who are starting today.

In the mentoring experience, you need to make sure you are providing the right information and advice. Mentoring really is about offering from a broader perspective advice on navigating the current environment.

My own mentoring is becoming more difficult, because the way we work is changing. Mentoring or advice is not about saying do this or do that. It is about opening people's minds about the opportunities in front of them. Or if they are stuck, giving them ideas about other paths.

Q. *Could you share any specific example?*

Let me give an example of a woman in one of our smaller offices in a remote location. This is one of the issues we have with small offices, because the expatriates develop a feeling that they are wedged with scarce opportunities, that they are seeing the same types of projects and not seeing the breadth of experiences that people in bigger centers get.

Generally, people in large organizations can grow networks. Most big companies have communities of practice that allow people to get involved in many ways and get skills beyond what they are doing locally. Short term assignments are becoming a more realistic opportunity. How can they build a path to where they want to go?

I let this young woman notice there are many other ways to develop skills in remote locations. Short-term assignments can reveal realistic opportunities for individuals in remote locations, so t they can figure out how to build a career path for where they want to go.

Q. *Have you mentored many people?*

I have mentored a dozen people during my 38 years with BP.

I never had assigned mentors. Any mentoring relationship I enjoyed was informal, because BP doesn't have assigned mentors. It happens "organically" in the sense that it is imbedded in the workflows of our daily routines, with senior personnel mentoring less experienced personnel, if there is willingness and affinity. I think organic mentoring is healthier than being assigned but is less likely to be successful.

With formal mentoring, there is a regular scheme of sessions, a regularity, as opposed to one-off encounters. In my perception and experience, single sessions can be useful for providing timely, specific advice. I would do that for any employee.

Mentoring is also about building a relationship, because I can see what happened to my mentees through the years, and can ask more pertinent questions such as, "Are you thinking about this, because you are replaying a situation that happened before?" or "Is that really relevant for you?"

I want to share one example. I have a female mentee, who I have been mentoring for 17 years. It has been great to see her career take shape and change through time. We don't have scheduled sessions. She calls me as needed when she has an issue. It has varied over the years, sometimes dinner or lunch. Now she is in Kuwait, so most of our contact is via phone. The contact style and especially the content of the mentoring has varied with time.

Q. *What can you share about your experience with career mentoring?*

Excellent question! I consider that career mentoring falls into multiple categories. I would like to highlight two aspects.

The first one is the business-related advice—How do you maneuver your way around your business? Then, there is the personal side—How do you maneuver around individuals? For example how to cope with someone with a very negative attitude. How you work with that kind of person can depend on the circumstances. A mentor needs to be open to these different aspects.

Q. *Recently, there have been movements pertinent to women raising their voice about mistreatment or harassment. Gender and culture have obvious implications in human relationships. Do you have any feedback on this?*

I did have an experience with a one-off mentoring. It was with a woman employee in my company, who recently resigned because she didn't feel her career was progressing properly. I never had a mentoring relationship with her. When I heard she'd left, I asked her if we could talk and it was very difficult conversation. She got very emotional and teary. In one word, uncomfortable. What would have been my appropriate response? My wife, a very empathic person, would have given her a hug, but I didn't feel comfortable doing that.

Q. *You have a long experience in the oil and gas industry. Was mentoring of women any different in the early times?*

I started with Sohio in Alaska, as part of a large intake of new hires in their twenties. It was quite diverse with a lot of women in reservoir simulation, in numbers about as high as today. Later, I realized how unique that was. In other environments later I felt something was missing. I have gone out of my way to have diversity in my teams.

Diversity of thought is important. To get diversity of thought, you need diversity in the team, not just diversity of experience. You can't build a team with diversity of thought without having a team in which each member looks different. Diversity is useless if you are not including everybody. Inclusion is just so basic, I tend to forget about it.

Q. Have you been vocal about diversity of thought?

Yes, definitely. When I have one-off single sessions with people and someone addresses an issue, then gender comes up.

For example, an engineer who looks for the trouble in things and is a little bit obstreperous, asking questions is good to have in moderation. Some make that approach their attitude, and people use different adjectives to describe them. If it's a man, they say, "He's stroppy." If it is a woman, they say, "She is a blocker." You see? Adjectives are used differently.

I have given many talks for BP and the UN about these topics. I ask the audience to think of a successful woman on their team and think of five adjectives to describe her. Then ask them to think of five adjectives for a successful man. Generally, adjectives used for men end up being technical, astute. For women, you usually hear team player. There are some evaluators or assessors who are more unbiased, and in their descriptions, can show both sides.

Q. Have they been mentored to be unbiased?

About 36 (maybe 37) years ago, in the early 1980's, times were getting tough and there were numerous layoffs. I remember my boss telling me, about the benefits of specialized professionals versus generalists and BP wanted generalists, but my boss called me in and told me, "If I was the best reservoir engineer, I would always have a job in this company." That was a very good advice.

I was and am a technical person at heart. That worked for me, but I don't think everyone can be a generalist or a specialist. Trying to fit into a box of what is trendy is not necessarily the right thing for everybody. I recently read how the US Navy is changing its warships from the earlier schemes in which each sailor would do one thing in specific. In the new ships, everybody can do everything. This is changing the way sailors work and the designs of ships.

I can see parallels of those changes with what is happening in business. We are all trying to be agile in many ways. It is a challenge for some specialist to work in that way. It simply is very hard to work in an environment where there is no quiet time to think. This article I was referring to showcases how this new approach has impacted the sailors and how ships are working.

All the sailors were doing something on board, but when there was a problem with the cooling system, there was no specialist aboard to fix it. The generalist approach can fail. We will always need those specialists with deep knowledge of specifics in our industry. We need to get the blend right, so that it can work.

Q. *Are we going through a digital transformation in oil and gas which may entail reverse mentoring?*

I don't like the phrase "reverse mentoring." When you are mentoring, you give skills or advice to others. It is not common to have young people mentoring older people, but we do have that. Our CEO has a young petro-physicist, who he asks advice about his own posts on social media. He is open about it, and tells us about it all the time. I personally don't have someone young who is giving me advice, but there are people who I go to for talking about certain things.

We definitively are changing the way we are working. We need to pay attention to the changing pace at which we are updating the tools around the company. Technology is always changing. Some of the stuff we were implementing some time ago has become much less expensive. We can run gigantic models quite easily and rapidly. Somethings are not necessarily new, but just faster.

Q. *Do you have your own Twitter, Instagram?*

No, but in BP I use the corporate tool Yammer, which is like a business-oriented Facebook within a company with lots of different channels.

With the change to digital and agile ways of working, I am not sure my advice will always be relevant, but I can raise questions. When organizations are structured differently, you run the risk of whether you can add value.

I do not have experience about how to evaluate the risk in mentoring in the future. In the UK, BP has a close relationship with the Imperial College and many ex-BP people approach this academic institution to work with students. I want to do something like that. Also, in the UN, I like to engage in mentoring people in developing nations. Those sorts of relationships will grow over time into new mentoring formats.

I remain concerned about the use of on-line systems, as it is hard to build the trust on-line.

Q. *Please tell us about your experience with sponsoring. Have you been sponsored to advance in your career? How? Have you sponsored others during your career?*

I had a couple of sponsors throughout my career. People who can move you up two levels in a short time frame by advocating for you.

Anyone can be a mentor, but not everybody can be a sponsor. To be a sponsor, you must have the ability to put words in the right ears to help people progress or you must have the power or authority to take the decisions yourself. You need that capability. Where I have done that is either with my direct employees or those I have mentored for many years. I do not sponsor everyone I mentor, only those people with the potential to advance.

To sponsor entails a level of risk.

Q. *Has anyone sponsored you in a critical way?*

I want to share two cases.

At one moment in my career, I had to move to a location that I was not particularly happy about due to the role I would have to do. I envisioned my previous manager, who trusted in me was a possible mentor or even sponsor for this situation. I called him and explained everything. In previous years, I was mentored by him, so he knew me well. I was offered a change and I realized the offer was a result of his action!

Maybe he wanted to bring me along with him. You do see this quite frequently. When people move, the move along with others they feel are their "team".

The second case was when I was moved into the job I have right now, where I look after BP's reserves. When the person who had this job prior to me had to retire, I had several people who sponsored me for this big role, which was a huge step for me in terms of accountability. I got it. Without a sponsor I wouldn't have gotten my current role, which I have had for nine years.

Q. *You say people sponsor someone who resembles them in some way.*

That is a real danger. I hope I do not do that, and I do not think that I do. One needs to be careful about that.

The people who sponsored me, look a lot like me. I don't know what were their reasons, but I certainly was the beneficiary of that. I like to think I don't do that. I do look for diversity.

Q. *In your own experience, what are the questions that the mentees should ask you?*

I don't want to be asked, "What do I do?" When people describe a situation, we may then discuss different options about how to progress.

Mentors should ask questions. That is part of mentors' responsibility. They must come forward with questions, so that mentees find solutions by themselves.

Q. *Do you think mentoring has to be in person or are the on-line and digital mentoring solutions equally effective?*

It is better face-to-face.

Q. *Please tell us about your own experience as mentee.*

My mentors are the people I can go to when I am in an odd situation about who to go to next, to explore options.

Recently I had a business issue that I knew was going to be sensitive. I have two bosses and I wanted to make sure they both were happy. In this case, I didn't know what to do. I wasn't sure how they would respond to something, so I called one of my old bosses from an earlier job position, and we got together for a coffee. I described the situation to him, and he asked some questions and gave me some ideas on how to approach the problem.

Q. *Is gender or age important in mentoring?*

No, I don't think so.

Q. *What do you wish you would have done differently in reference to mentoring or sponsoring, as a mentor or as a mentee?*

I would have gotten more out of my career if I had been more active in searching out mentors or sponsors earlier. It is hard to seek out a sponsor. People need to be seeking out good mentors.

Q. *Do you seek sponsors, or do sponsors come forward and sponsor you for some reason?*

In my case, it was affinity. There is a lot to do with my alma mater, too.

Q. *Did you enjoy any mentoring/sponsoring experience? Why? Tell us more about those joyful occasions!*

What comes often to my memory is an anecdote about when I was working as an expatriate and I had mentored a young person there. When I returned twenty years later as an SPE speaker, that person was the keynote speaker. He had a high position at the Bureau of Mines. I was so glad to see his success. I remember that person coming in on roller skates and wearing a bowtie in the office. Things change.

A Shared Selfie

- **Your favorite role model**: Pete Goss (a well-known British sailor).
- **One word to define your experience with mentoring**: Relationship.
- **One word to define your experience with sponsoring**: Caution.

- **Who you wanted to mentor or sponsor you, but you never had the chance to ask?** Dietrich Bonhoeffer (a German theologian, who was involved in a plot to assassinate Hitler).
- **What question would you have asked this mentor?** What's the right thing to do that is not always right? This question explores the ramifications of his decision. He must have considered technical, legal and other factors.

Afterthought

David told us he was about to retire, and he was getting prepared for his post-business life, by thinking about becoming a sailor and planning to do a lot of offshore sailing.

His answer about a role model triggered our attention. So, we asked. This is his answer:

> As I am getting ready for retirement, I met Pete Goss. He has provided me with loads of advice on preparing my own boat and getting ready. This mentoring experience taught me that it is important not to be afraid to ask someone to mentor you and that you should not be afraid to ask anyone. Some people really want to share their own experience. Don't be afraid to ask someone to help.

The Other Side of the Coin: The Mentees

"You have to be open to opportunity."

Magy Louise Avedissian.

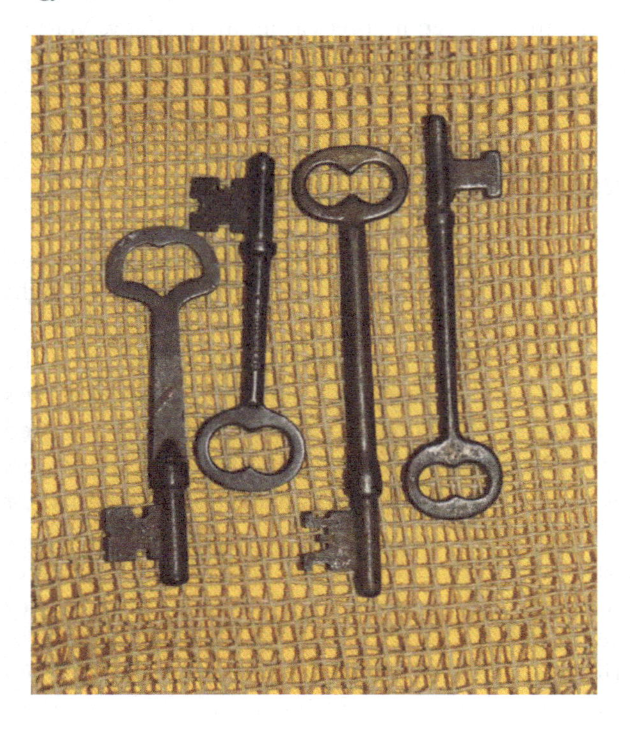

M. A. Capello and E. Sprunt, *Mentoring and Sponsoring*,
https://doi.org/10.1007/978-3-030-59433-6_26

Interviewing Emerging Professionals

To identify what mentees want and expect to get from mentoring, we interviewed some young professionals and emerging leaders. Our selection was based on their involvement with the Society of Petroleum Engineers (SPE) and the Society of Exploration Geophysicists (SEG) and the recognition they have received from those groups, especially awards and leadership roles.

We originally planned to include separate chapters for each interview with one of these outstanding young professionals with one to seven years of experience. However, we found most of their responses about mentoring were very similar.

There were some striking differences, mostly related to sponsoring. The nature of their employment (e.g., corporate or academic) and the culture in which they are working has a significant impact on sponsoring relationships.

The motivation and understanding of the importance of mentoring for career advancement varied, depending on their cultural context. If the emerging professional was shifted from his or her homeland and immersed in a new cultural context, mentors and sponsors were even more important.

We used the same questionnaire to capture their insights. To ensure we heard each mentee's unbiased perspective, the interviews were conducted separately. We will refer to the interviewed mentees as Jennifer, Magy, Matteo, Ruth and Wassim. Portions of their answers are included verbatim to illustrate specific points and perspectives.

"I Am in Charge" Perspective

The young professionals we interviewed appeared to be actively in charge of their mentoring. They seemed to have sought mentors when they thought they would be of value.

In contrast, many of the older leaders we interviewed did not realize the value of mentoring until later in their careers. Back before everyone was talking about mentoring and sponsoring, mentoring and sponsoring occurred spontaneously and some leaders naturally benefitted. In hindsight, the leaders recognized the important role these advisors and advocates played in their lives.

The early-career professionals were not shy about declaring their awareness of the value of mentoring. They actively pursued mentoring by key individuals as much as they could.

Remarkably, many of them also realized the importance of having a variety of mentors. Some of them may be appointed to higher roles or leave their companies to pursue other objectives. They are aware the work environment is very dynamic, and we noticed an interest in obtaining the most of a mentoring experience as rapidly as feasible.

Role Models

The role models of the younger professionals are very different. This may be due to generational differences. Younger people have a broader concept of success and their goals are shaped by social media influencers and tycoons who are prominent on the internet.

Also, the reputations of some former role models and leaders have been destroyed by movements like *#MeToo*. Prominent and powerful men, who have abused and taken advantage of women, have been disgraced.

The internet is filled with stories of humanitarian leaders of all ages and nationalities who are actively engaged in fights to protect the environment or eliminate violence against women. Some of these role models are still in their teens.

Importance of Sponsors

Another key finding is that the value of sponsoring is not widely recognized at this early stage of their careers. Sponsorship makes a huge difference in the trajectory of careers, influencing not only the direction of the career, but also someone's ultimate success. As we move through life, each step is strongly influenced by prior experiences and progress. Development and advancement are cumulative. Experienced people often regret not realizing the importance of sponsoring sooner, and the younger ones are often not aware of it.

Early-career professionals often think their good performance will speak for itself. They believe their accomplishments and expertise will be noticed and rewarded. From good and bad experiences, older professionals recognize the huge difference sponsors make in shaping and accelerating careers.

We shared our insights about the relevance of sponsoring with all of them. Unfortunately, many of them will probably wait for their good work to bring them sponsorship.

Many people in minority groups make this mistake. This is the case of minority ethnic groups such as Latinas in the USA or Europe. Many women also suffer from this mistake, waiting for their performance to shine and bring them promotions. Younger men may not be more proactive, but they tend to benefit more frequently from sponsorship, because current male leaders tend to recognize younger men as their youthful counterparts and sponsor them.

Unfortunately, the #MeToo movement may make some men more reluctant to sponsor women and/or to meet with them privately. Men may fear that unscrupulous women, who have been disappointed, will try to damage their reputations. Young women may wish to arrange meetings with potential sponsors in public places or with multiple people present.

We hope this compilation will help in raising a necessary awareness related to mentoring and sponsoring. There is a natural tendency after doing well and being recognized in school to believe that the same process occurs throughout one's career. It doesn't. We all need as many sponsors as we can get.

Below, we share with you in their own words the ideas of our early career professionals.

Q. Tell us about your own mentoring experiences

Wassim: I am a Moroccan computer scientist, and my role model is the boss of my boss at the company for which I work, because he showed me a completely different perspective on the work. He role-modeled sheer excitement and humility. He didn't use his massive knowledge to intimidate others. He explained that he was still learning himself and that at no point in life should you stop learning.

Magy: I am a Petroleum Engineer of Armenian origin. Mentoring has definitively been a cornerstone for me. I have had a pragmatic approach, pursuing internships to receive practical advice about my way forward. Then, I sought mentoring from experienced people in academia and in oil and gas to get a broader outlook into my future. Given my humanitarian interest for my career, this approach helped me find my current job in Houston and my decision to pursue a MSc in Petroleum Economics

Jennifer: I am an accountant who studied in France. I have worked for a consulting firm, first in the UK for five years, and now in Texas, USA. In my company, there is a structured mentoring program for women and minorities, and I have benefitted from it, but I also seek mentors outside my company. As I spend more time working on my career and get to know people outside my company, it is valuable for me to receive guidance about my own career from them.

Matteo: I am an Italian geophysicist and studied for my Ph.D. at the University of Edinburgh. I found it much easier to get a mentor in the UK than in Italy, because the educational systems are very different. Most of the time, I do not need mentoring for technical matters. I want mentoring for career advice, such as whether to stay in a technical area or move into a managerial role. For this, I seek mentoring outside of my company because they may be less biased. They could be more focused on my specific problem.

Q. In your own experience, what are the questions that the mentees should ask?

Ruth: I am an American computer scientist. Before asking questions, I think it is important and fundamental to establish a trusting relationship. My favorite questions are "What can I do better? What are my strengths outside of my job function?" I need to apply an emotional intelligence perspective. I am always seeking and thinking to whom should I talk next? How will I get to know my next mentor?

Magy: The things on which I always concentrate are the challenges a person faces. So, I ask about the challenges and things I should avoid.

Jennifer: I think the questions come due to preparing for mentoring meetings, with lots of self-reflection about where you are in your career and what it is that you want to get out of that mentoring. When I was seeking to be promoted to a manager role, there were many different angles. For me, it was more about asking specific questions. My questions were about change. I would share with my mentors what I had prepared, and ask, "I have done this and that. What do you think I need to do in addition?" At the end, I think you must be motivated about how to progress your career. If you have that desire, your ambitions should drive the questions. The right questions and motivation are important.

Q. Do **you** ***consider that single sessions, or mini-mentoring sessions are also effective?***

Ruth: Mini mentoring is most effective. Most mentors don't have a lot of time. I had to be prepared. My mentor didn't have a lot of time. My preparation before meetings was important.

Wassim: I prefer continuity.

Magy: Those single mini sessions are so vital! You never know what you will learn from those mini-mentoring sessions. They can resonate with you.

Jennifer: For me, the best mentoring sessions I have had were out of the office, over coffee or lunch. Short and relaxed. Even the environment is important. You must feel comfortable to be able to ask open and honest questions.

Matteo: I have never had a structured mentoring program, so I do not know if a program of mentoring would be good or not for me. When you need career advice, stand-alone mentoring works. I have discussions with my colleagues that I consider to be peer-to-peer mentoring on how to approach problems. They are a one-hour discussion with more experienced people on specific issues they know how to handle. You can save time with punctual, single sessions.

Q. Did you* have *a sponsor?

Wassim: I had endorsement by peers. I have not yet had a sponsor.

Matteo: I guess this is very hard to say, because you are never 100% sure. I guess that when I received the SEG J. Clarence Karcher Award (2018), I had to have a nominator, a sponsor.

Magy: I really do not know. There have been instances in my life where people would speak well about me.

Jennifer: Yes, I have a sponsor. It was a result of a mentoring program in my company. Some mentoring relations evolve into sponsoring. Once you get a sponsor based on trust, you have been given an opportunity. It is then up to you to make the most out of it.

Q. Do you consider* mentoring *has to be in person, or are the on-line, digital mentoring solutions equally effective?

Ruth: Mentoring in person is best. I do not think social media mentoring exists.

Wassim: Social media-based mentoring? Does that even exist? I never received mentoring remotely. I definitively prefer in person so that I get eye contact and can see the mentor's facial expressions.

Magy: We live in the digital age. If you feel on-line mentoring is not effective, you are missing out on opportunities. You must be open to opportunity. I had a mentor in The Netherlands from Shell, who helped me a lot, always remotely and others in France. I am very keen on receiving mentoring remotely.

Jennifer: I haven't experienced mentoring other than in person. On Instagram and LinkedIn, I am seeing more and more inspirational quotes, especially about women in industry, nominating women for awards, and the scarcity of female awardees. There are many women role models who are inspirational. The difficulty resides in understanding how the quotes relates to me.

Matteo: I do mentor "moments," and mostly these are on-line. When it comes to career advice or anything related that is not purely technical, I do not see any issues with being on-line with someone who will mentor me.

Q. Is gender or age important in mentoring? Females mentoring females, or older people mentoring young? How about for sponsoring?

Wassim: Culturally, I cannot challenge someone older than me.

Magy: No, it is not important. Honestly and granted, I have had a lot of powerful women role models and will add you two to the list. In retrospect, I have had life-changing advice from male mentors. I will be candid in sharing that in my opinion, we are placing too much emphasis on the empowerment sphere. Mentorship should emphasize wisdom.

Jennifer: Experience is priceless, so I think it is valuable to have an experienced professional mentor younger or more inexperienced ones. As far as gender, it is a very interesting topic for me, as all my mentors have been males. I never had a female option as a mentor, as in my company there were no female partners in the practice of oil and gas. Role models are very important. My male mentors have been exceptional and not treated me any different than they would a male mentee. I have never had a negative or uncomfortable experience in mentoring due to age or gender.

Matteo: For me is not important, but in engineering and in sciences, where women are in the minority, there are questions perhaps that are best discussed between women.

Q. Who mentored/sponsored you in a memorable way? Why do you consider that was memorable?

Matteo: If you mean unexpected, it was probably the (unknown) sponsor who nominated me for my SEG award.

Q. What do you wish you would have done differently in reference to mentoring, as a mentee?

Ruth: I wish that I had said thank you more often and made my appreciation known to my mentors.

Wassim: I regret forgetting to say thank you explicitly for all the many ways in which the input of my mentors was valuable. In Morocco, people are not very direct about feelings—I want to have a chance to say thank you.

Magy: I don't have regrets. I asked the hard questions I needed to ask. I'm happy.

Jennifer: There is probably more preparation I could have done in terms of self-reflection before the mentoring meetings.

Matteo: Sometimes, I could have benefitted from asking a mentor my questions earlier, but I waited too long because I thought I could solve the issue myself. I may have been too shy, or I may have mistakenly thought the issue was much more straight forward. Now, I realize many issues are multifaceted.

Shared Selfies

We were surprised our young professionals did not share similar role models, and that the words they chose to relate to mentoring and sponsoring were so dissimilar from one another.

We tend to cluster young people in one big bucket, but now more than ever before, younger professionals develop personal perspectives, perhaps driven by the overwhelming flow of digital media, in which they are fully immersed and thrive. In sharing this section, we hope you will gain insight into why young people should be evaluated as individuals and not primarily as members of a demographic group.

- **Your favorite role model**:
  - **Ruth**: Shonda Rhimes[1]
  - **Wassim**: Aaron, my current supervisor
  - **Magy**: My mother
  - **Jennifer**: I am still thinking about it. It still must come
  - **Matteo**: My father.

[1] A TV writer and executive producer of "Scandal" and "Grey's Anatomy" .

- **One word to define your experience with mentoring:**
  - **Ruth**: Reflective
  - **Wassim**: Learning
  - **Magy**: Wisdom
  - **Jennifer**: Supportive
  - **Matteo**: Direct, honest.
- **Do you want to share the name of your most important mentor?**
  - **Ruth**: Neal Githens (the boss of her boss, *verbatim*)
  - **Wassim**: Aaron, my current supervisor
  - **Magy**: My mother, Margherita
  - **Jennifer**: A partner in our London Office.
- **Who would be an ideal mentor for you in this moment of your life?**
  - **Ruth**: Sheryl Sandberg (Facebook COO)
  - **Wassim**: Sam Harris (a philosopher and cognitive geoscientist) and President Barack Obama
  - **Magy**: Charles Aznavour (a French-Armenian singer who has received many awards), because he uses his craft to help people
  - **Jennifer**: Oprah, because she collected answers from thousands of people.
  - **Matteo**: Steve Jobs, because he was a visionary.
- **What would you ask this person?**
  - **Ruth**: Does it get easier?
  - **Wassim**: How do you keep calm and composed in the face of so much negativity and criticism?
  - **Magy**: How do I use my craft to help my homeland, Armenia?
  - **Jennifer**: What was your biggest fear? Are you still dealing with it?
  - **Matteo**: When and how did you realize that something has potential something and is worthwhile before doing it?

Afterthoughts

The conversations we held with these early-career professionals triggered some questions we would like to share:

- Are role models losing value or even relevance for the new generation?
- Who are current role models? Who will be role models in another decade?

- How can on-line mentoring processes be effective across multiple cultures to promote rapid development of new leaders?
- How can on-line processes preserve important experiences and insights of retiring professionals?
- Can a virtual mentor be constructed by combining characteristics of multiple experienced people with machine learning and artificial intelligence?

Difficulties in Sponsoring and Mentoring

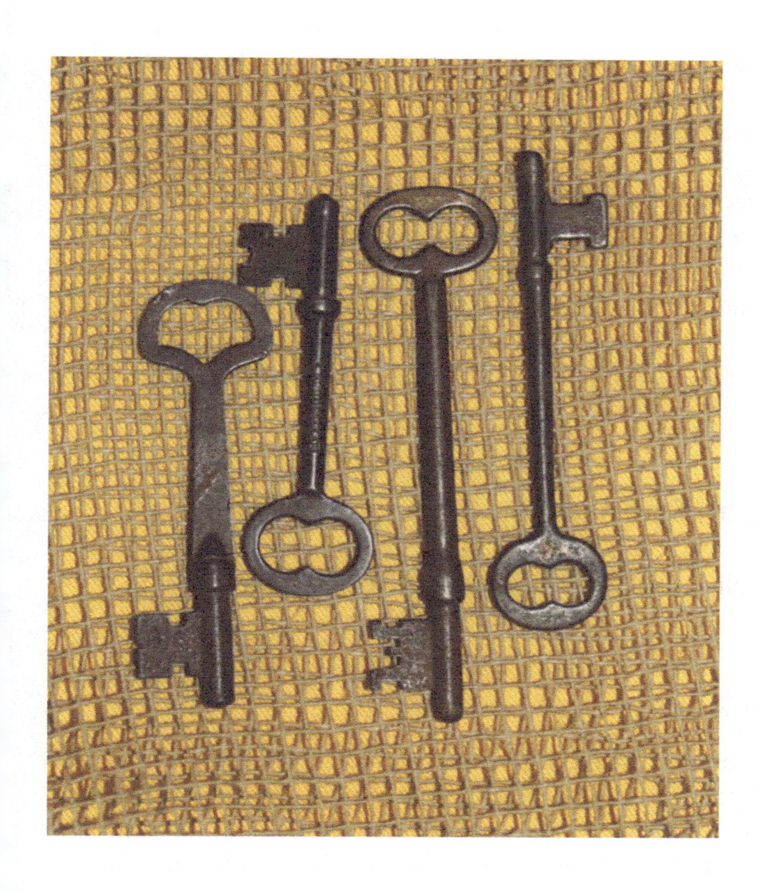

M. A. Capello and E. Sprunt, *Mentoring and Sponsoring*,
https://doi.org/10.1007/978-3-030-59433-6_27

Not all sponsoring stories are success stories. Sometimes an individual may experience the downside of sponsoring, which could include:

- **Loss of sponsors**
- **Unrealized sponsoring opportunities**
- **Sponsors who engage in harassment.**

Many times, immersed in the advancement of their careers, individuals trust their sponsors will open opportunities for them, and don't build and maintain a strong network, or find multiple sponsors inside and outside their organizations. Also, not all sponsors behave honorably. Some sponsors may overstep their role or abuse it, and unfortunately engage in unethical behavior.

Women in particular face additional difficulties obtaining sponsors and mentors. We provide some examples in the following sections, for your own self-reflections.

We realize not all people who experienced negative experiences with sponsoring were willing to disclose them. Individuals who experienced harassment may not be at freedom to talk about it, or choose not to disclose the specifics.

So, we recognized this will be a topic of another research or compilation of insights, as there is much to learn about how to avoid losing sponsors and avoid abuse and harassment in professional contexts.

To share some of the potential pitfalls of sponsoring, we are sharing in this chapter some cases, including our own, of professionals who experienced some of the downsides of sponsoring.

Missing Out on Sponsoring Opportunities

Melody Mitchell, Lieutenant Colonel (Retired) in the US Air Force

Q. How would you describe yourself?

I am a leader, a newly retired Lieutenant Colonel (Lt Col) in the US Air Force. I am the leader and founder of Lean In Kuwait, a wife, a daughter, and a sister. In my job, I had the opportunity to sell bombs and planes to Kuwait military. I mediated between the needs of the country and helped them to understand their requirements and convey those needs in the USA.

I am a diplomat and I worked in the U.S. Embassy in Kuwait. I am racially ambiguous, and I tend to fit into different communities, because I blend easily. I take the opportunities to explore different communities and to learn as much as I can. I am a student and always learning.

Q. *Who were your main mentors?*

That is a tough question for me. Earlier in my career I did not have mentors. There was a Lt Colonel detachment commander in Iraq. I met him when I was a Captain reviewing his unit and really admired him. I boldly sent him an email asking if he would like to be my mentor. When he did not respond, I felt embarrassed. Even now, I do not know why he did not reply. This experience made me too shy to seek other mentors.

Around the same time another Lt Colonel offered to mentor me; however, he first tasked me to get a list of my peers so I can "compare myself to them and see where I rank." It made me uncomfortable and I did not complete his request nor follow-up with him. I would rather focus on me and what I need to improve versus compare myself to others.

There are two books that were life changers for me, but I did not read them until halfway through my career. I started with the Air Force in 2000 and around 2013 I read "*Lean In: Women, Work, and the Will to Succeed*" by Sheryl Sandberg. Shortly afterwards, I also read "*The Charisma Myth*" by Olivia Fox Cabane. Those two books opened by eyes to networking. Looking back, I realized I missed out on many mentoring and sponsoring opportunities due to lack of technique and awareness.

From Lean In, I learned I should not directly ask someone to be my mentor and slowly develop the relationship through questions. I also realized I had the Tiara Syndrome.[1] I just thought that if I do my job well, eventually I would be recognized for my hard work. I networked well with my peers but did not convey myself in the best way to the ranks above me. I decided they were too busy and just focused on my job.

What I learned from Cabane's book is that my supervisors want to help and mentor you. By asking them for help, they decide to invest in you. My approach was that I could prove I could do the job without asking clarifications or questions. Therefore, I never asked for help, because I considered that showed vulnerability or weakness on my side. So, I missed out on mentorship opportunities.

[1]Tiara Syndrome is a term widely used for exemplifying the lack of willingness to seek sponsors by women, as they are generally convinced their good work will be noticed and someone will crown them with a tiara, meaning a promotion, a salary increase or an appointment. This almost never happens.

After reading those books, I was able to develop strong mentors and sponsors.

***Q.** With regards to sponsoring, could you tell us your experience seeking sponsors and any best practices or lessons learned?*

I missed an opportunity. As a Captain, I built a relationship with an Air Force Colonel. He worked as our region commander. In my work, I had the opportunity to interact with him extensively and developed a lot of respect for him. I could have pushed that relationship further into sponsoring and mentoring by asking more questions and reaching out, but I did not. In time, he became a General and commander of our entire organization. Others enhanced their connection with him, but I pulled back, so as not to bother him in his new important role.

A couple of years later, I had a challenge with my chain of command and he was still the highest-ranking leader of our organization. I could have reached out and asked him for help but I did not. Instead of defending myself, I felt shame as if I let him down and did not reach out to him at all. I knew he did not have all of the information, but I did not control my narrative. The Brigadier General would have been a great sponsor. It was my own doing that prevented that from happening.

You must invest in all relationships at work, even if uncomfortable with bosses, push yourself outside of your comfort zone and invest in those relationships. Find various reasons to stay in touch and maintain contact as you both continue in your careers.

***Q.** Are there enough women in mentoring roles in the military?*

There are two parts to that answer. Yes, we do, but we still need more representation of women in our ranks. We are 20.9% women in the Air Force, and there are only 13.9% in the senior rank of Colonel. We definitively don't have enough women Generals with only 7.9%. We have had only three 4-star woman Generals in the Air Force (the highest rank).

One of the things that I feel women in the military need is support and mentoring from and at multiple levels. Many times, we are the only woman in the room. We are very diverse in thought, so when we see another woman, that does not mean she will share the same point of view.

In 2013, a friend added me to a Facebook group called "USAF Women Officers Forum," an unofficial group for women officers. I immediately reached out to the admin and asked her if I could assist. Since then, we added two more admins and the Facebook group has grown to almost 11,000 women in the global network. I consider it an affinity group and an employee resource group. I feel it is very special. Women ask questions supporting all aspects of their lives, openly or anonymously.

Q. Does this Facebook group act as a collective mentoring group?

Absolutely! We feel it provides the opportunity to reach out, with many women offering information and insights. Numerous mentoring relationships have developed. A photo album called "Friends from the Forum" highlights women officers who became friends by meeting in the forum and later meeting in person. The forum has created many cherished relationships: friendships, mentorships, and sponsorships.

Q. What would you recommend to the new professionals in your sector or in general?

What changed for me after I read those two books was so huge, that I recommend those books to everyone, especially to women. When you are so focused in your work and your duties, you may miss out on professional networking opportunities. Networking with other women at a Women In Defense event led me to Lean In.[2] Lean In recommends peer-mentoring and provides a framework of self-empowerment that I find highly beneficial. I could have used that information earlier, but it was not too late in my career.

I would challenge people to be bold enough to reach out to sponsors and mentors and groom the relationships with questions and updates to champion your progression. I also challenge them to seek peer-to-peer mentoring. I recommend to absolutely and immediately start to build all those types of relationships. Don't dismay. Don't let other people's actions stop you from trying. Cast a wide net and if you miss in one way, try with another style.

Misinterpreting Sponsors Subtle Inquiries

Ken Tubman, Former Executive in a Variety of Operating Oil and Services Companies

Q. How would you describe yourself?

For interests, I am fascinated by technology and especially by the implementation of technology to create value. Too often people forget the value aspects and don't go beyond the search for an interesting piece of technology. About

[2]Lean In is an initiative launched by Sheryl Sandberg, COO of Facebook, to empower women. There are 44,000 Lean In Circles in over 170 countries.

my style: I usually ask a lot of questions, and I can be outspoken and blunt. I realize that has resulted in a reputation for being tough. Friends have pointed out, and I need to acknowledge, that reputation is not fully accurate. I truly enjoy hearing other people's opinions, debating, and having those debates shape my views. I consider myself open to new ideas and I was quite surprised the first time I heard that my questioning style gave the opposite impression. I also enjoy working with and helping others. I very much enjoy helping others think through challenges and develop their capabilities.

Q. *Tell us about your own experience with sponsoring and mentoring.*

There have been multiple people whose opinion and guidance I valued and that helped me tremendously. I am still appreciative of the time and wisdom they shared with me. It was never in a formal "mentoring session", but there were many great conversations.

I have been very happy later in my career to be able to return the favor with others. Some of my favorite interactions have been with younger stars cycling through our group on development assignments. These people would come in from different disciplines with very different backgrounds and experiences. It was terrific seeing how they approached problems and added to a team of very experienced professionals.

I do not remember having many sponsors, perhaps maybe one. I do try to act as a sponsor for others, when I can. When I use the word "sponsor", I mean someone who was proactively pushing me along in my career. That's different from a mentor who is providing advice and counseling. There is an example or two I can think of, but perhaps there were more behind the scenes?

Q. *Can you share on example?*

One example of a mentor helping me was at Veritas DGC. The president called me to talk about a different position. I don't remember the exact conversation but I do remember I was out of town so we were talking on the phone. It was a terrific position but what I remember of the conversation was that he seemed to be just probing, suggesting that I wouldn't be interested. I assumed he was just letting me down easy, giving me a bit of advance notice that I would not be offered that role. The next day our head of HR, someone I am very proud to consider a friend and mentor, called me. Knowing me well, he was comfortable being blunt. Again, I don't remember the exact words, but it was something along the lines of "What the hell are you doing telling the President you won't do the job he wants you to do?" I was in shock at how badly I had misinterpreted the first conversation. Fortunately, I had someone who pointed out my mistake. If not, I would have inadvertently turned down the best job I ever had.

That same mentor continuously provided nuggets of guidance. From small tips to strengthen executive presence to brief comments that struck right to the heart of a behavior I needed to modify. I also remember times when he would be very subtle in testing my progress on one of those points.

One example of sponsorship that comes to mind was with one of the people who had a rotational assignment in our group. It was a subsurface group and his background was mostly in operations. The main point of the position was to give those from other disciplines background in subsurface. Sure, it's not at the same depth as someone doing interpretation or reservoir modeling, but it certainly gave broad exposure to the key issues and different approaches. Even without the ability to solve the problems themselves, people who had been through the position would know the right questions to ask and the right people to call for assistance. In a talent meeting there was a discussion of different people's experience. It turned out one leader was challenging that this person had any knowledge of subsurface. Fortunately, I was in the room and we were able to talk through it. To me that's an example of sponsorship that is much more subtle than overtly moving someone into a new position. In this case, the person was clearly strong enough to earn their own opportunities. The benefit was keeping the record straight and removing potential barriers.

Q. What other advice is relevant in relation to sponsoring?

One concern I have about sponsorship is the impression that a new position depends on that sponsor. Having a supporter is great. Depending on someone else to get you another opportunity is not a good approach. Sponsorship can help open up possibilities but I caution people to continue seeking and generating opportunities on their own.

One time I had conversation with a long-term employee at my company. He was expressing concern that all his sponsors had left the company. I found this surprising. Not that people had left the company but that I sensed a feeling that without those sponsors no new possibilities would come forward. Since I have changed companies a number of times, I always felt it was up to me to perform well enough so opportunities would be available.

Q. What would you recommend about mentoring?

I believe the best mentoring relationships develop naturally. It is very important for both sides to be comfortable in the conversations. And it's more than just the conversation. It truly is a relationship. I believe the mentor needs to stay in touch and keep pushing. The objective is not to just throw out pearls of wisdom and feel satisfied. The real objective to help the mentee grow and develop. That means checking in, and testing progress. There have been times where I will talk about course of action with someone. I often feel it is also my responsibility to follow up, not just theirs. I can't force them to act, but I can help with reminders and nudges. It will make for a much more successful relationship.

I had no exposure to formal mentoring programs until fairly late in my career. Honestly, most of my experience with these programs felt contrived, and were not successful. Of the multiple times I was "assigned" a mentee only one or two turned into what I would consider a helpful relationship. And one of those lasted only for a few conversations.

I would draw a contrast to relationships that developed from working together or some other common interest. A level of trust built first, and only then did the conversations become deeper and meaningful. These are the relationships that make for good mentoring.

Q. What would you have changed in relation to mentoring or sponsoring for yourself, now looking your career in retrospection?

One thing I would emphasize is the need to maintain the relationships. As both sides move around to different groups or different companies it can take extra effort to stay in touch. That effort is clearly worth the investment. In hindsight, I wish I had done better with that.

That need to stay connected goes both ways between the mentor and the mentee. If it is a relationship worth sustaining then it is not fair to put all the responsibility on only one side.

Not Noting Signals

Emma Perfect, Executive Assistant in a Consulting Firm, Middle East

If the mentee or sponsored individual is in the midst of difficult situations such as a downsizing and/or reorganization, indications of ulterior motives of mentors and sponsors may be overlooked or misinterpreted. The stressed individual may mistakenly think someone who is acting as a mentor will also be a "savior" and not notice signals that the relationship is going astray. The following example highlights how failure to recognize signals that a mentoring or sponsoring relationship is not purely professional can lead to problems.

We interviewed Emma Perfect, to understand how she coped with a mentorship/sponsorship relationship at work that went sour affecting her in several ways.

Ms. Perfect worked as an Executive Assistant in a consulting firm in a country in the Middle East. Her role enabled her to be in close and frequent contact with several of the executives of the firm within her local office and also in other countries in the region and around the world. At the time of the events, Emma was working her get her MBA to enhance her career opportunities.

One of the executives from abroad became a frequent liaison. He solicited her services and praised her work. He became friendly with Emma, and offered to help her with the assignments of her MBA program. Emma would remain at her office after hours, because it was quiet and she needed to study many documents and work on her assignments. He would also linger in the office to help her. She considered this a sincere effort to support her, because he was 20 years or more older than she was and married. She never imagined the relationship between them was more than strictly professional.

Emma's current boss was scheduled to retire and "the mentor" advised that she should start actively seeking another role within the company they both worked in. This triggered an unprecedented level of stress for Emma, as together with the challenge and commitment required to pursue an MBA while working full-time, she would now have the additional struggle of searching and then starting new employment. Not to mention the financial difficulties this could present, as she was funding her own tuition. Soon the "mentor" offered her a position as his Executive Assistant in his country. This appeared to be an appealing promotion for Emma. Unaware of any illicit intentions, Emma Perfect accepted increasing support from her "mentor" as she became convinced that this was an opportunity to avoid losing her job, ability to financially continue and complete her MBA and moreover, an opportunity to gain greater visibility within the company.

In retrospective, she remembered that some female colleagues from the Middle East confided in her that they felt uneasy with this fellow, because he touched their arms during salutations and greetings "a little longer than normal." However, she disregarded these observations, assuming that these concerns were because the other women were natives of the Middle East, with a different culture. For her as an English woman, the physical contact on greetings was not abnormal and he was her mentor. She didn't imagine the relationship would go wrong.

After a few weeks, Emma's mentor invited her to get to know the offices of the company in his country by taking a business-related trip during which she could "assess her new role and her new teammates on site." During the visit, nothing unanticipated happened. Emma assessed the conditions of the offer versus the cost of living in the other country, and declined the option, because it was not sufficiently economically attractive to undertake such a big change.

Soon after, during a business visit to Emma's country, this mentor/sponsor asked her for a ride home after working hours when it was quite late. This was the first time Emma received such a request, but she accepted. While she was parking, he forced a kiss on her.

Emma was shocked. She began babbling anything to get him out of the car. "*What is this? You are married. I cannot engage with you. Leave me alone now.*" She never saw it coming. She went home crying, feeling foolish and blaming herself for not noticing indications that his intentions were anything but strictly professional.

Perhaps, the issue Emma experienced was not so much the fact per se, but her own reaction, because afterward, she had difficulty coping with the event. She started to develop negative thoughts about herself. She could not understand how she became convinced this man was interested genuinely in progressing her career and hleping her achieve goal to successfully complete her MBA program when he was interested only in a passion relationship. This damaged her self-esteem. She said, "I tried to become *invisible by wearing clothes that disguised my body. I stopped wearing make-up at work and never wore my hair down, always tied my hair back. I even went to the point of dying my blond hair brown, worrying men would only look at me for my beauty and not my capability at work.*"

She changed jobs and started to work at a bank. After four months at the bank, her boss another female employee take her aside to advise her politely that she should try and to wear a little makeup and dress smarter at the office, the way she had looked at the initial job interview, before the incident that triggered her look-change.

This conversation and the support of her new boss, which was completely professional in approach and interest, enabled Emma to recover her self-esteemed and accept that the incident was not at all her fault, but a combination of not correctly reading the signs that were there and unethical behaviour from the supposed mentor.

It was not an easy recovery, as it took Emma one and a half years to regain her confidence back. She mentioned to us *"in hindsight, I realize that in light of possible pending unemployment and the immense pressure to complete the MBA program, I became absolutely desperate for support. So maybe I didn't read the signs or just brushed my colleague's observations aside."*

She advises any young person to maintain a rich network at work, paying attention to their feedback on colleagues. Different perspectives on behavior may significantly raise your awareness on the transparency and ethical profiles of colleagues that you may be considering as mentors or sponsors. Ms. Perfect also mentioned that boosting your self-confidence is pivotal to grow your career.

A Couple of Anecdotes from Eve

Emma's experience reminded Eve of her early career when she always thought that men's intentions were strictly honorable. Here are her recollections of two anecdotes, related to preventing harassment.

I got married when I was 21 and was working on my master's degree. My husband was suspicious of all my male contacts. When I traveled, he would have flowers sent to my hotel room. I told him that was a silly waste of money, because all I had time to do in the room was wash and get to sleep. Then at a conference an old colleague from graduate school came up to my room supposedly for me to get him some information. My husband told me that he suspected that some man would come up with an excuse and that the flowers were his "virtual presence" and a reminder to the intruder that I was not available.

The other event occurred when I went to a week-long Gordon Conference, which was a small workshop for 100 participants at a boarding school in New Hampshire during the summer. There were sessions in the morning and afternoon, but the afternoons were open for casual interaction. At lunch the first day, I tried to recruit a group to go for a ride or a hike and found only one taker—a French research associate, who had come to Stanford after I left. We went out for a drive and he indicated that he wanted to find a place to pull over. I thought that he needed to "relieve himself" and looked for a rest area with toilets. When I found one, he explained that was not what he wanted. On the drive back to the conference site, we had a long discussion. I explained I was married and did not engage in extramarital activities. He replied that he was married, but that his wife didn't care what he did. I emphasized that while it might be OK with her, it wasn't with me. It was a very awkward and seemingly very long drive back.

There were only four other women at the conference, but by the next day it was obvious that he had found one of them, who was perfectly happy to have a relationship with him for the week. During the rest of my career, I saw this man at various meetings, but I could never forget his "propositioning me."

Mentoring and Sponsoring of Women

M. A. Capello and E. Sprunt, *Mentoring and Sponsoring*,
https://doi.org/10.1007/978-3-030-59433-6_28

Our collection of a wide variety of personal insights from leaders and young professionals on mentoring and sponsoring has given us the opportunity to compare and contrast many people's perceptions about mentoring and sponsoring.

We have mentored and sponsored many people through the course of our long careers, and also received valuable mentoring and sponsoring for our own career progression.

Perhaps, our stories can help the readers to gain insight on women's experiences reaching out or receiving mentoring and sponsoring.

To add to what we have learned from our interviews, we'd like to share some anecdotes and elements of our own journeys to further illustrate facets of these two pivotal career enhancers. Since our experiences have been different, to enhance clarity, we will do this in the third person.

Our Mentoring

Maria Angela is sought after for mentoring, because she has a natural ability to inspire trust, and has experience in detecting gaps in leadership style and communication. In Latin America and more recently in the Middle East, Maria Angela has developed expertise in mentoring C-suite leaders and at other extreme of the career progression, new hires. She prefers to mentor leaders in one-to-one settings and new hires in groups including making presentations to large audiences on mentoring topics for new hires. She has expertise in mentoring across cultural barriers, technical disciplines, and particularly across different corporate environments, for example IOCs versus, NOCs (international versus national oil company culture), and academia versus industry. She has become a "translator" and finds it to be useful for mentoring across industries, and especially to give or receive mentoring in academia or in a corporation.

Although, she does not think gender was a major barrier for her career, Maria Angela acknowledges there are some useful tips for women to consider in relation to mentoring, and those are:

- To actively seek out mentoring.
- To actively expand your networks.
- Do not shy away of revealing your drivers and ambitions to your mentors. This will help them provide customized career advice.
- To ensure the availability of mentors for you within but also outside your organization.

- If you are entrepreneur, seek out mentoring from peers both in your own and other industrial/activity sectors.
- Keep the mentoring relationship at a professional level. Mentors are not necessarily your friends.

Maria Angela believes storytelling is essential to mentoring.

Eve has mentored many people in a variety of ways. She prefers lean, fit-for-purpose mentoring to formal, structured mentoring programs, and when she served as a mentor in highly structured programs, whether within her company or through professional groups, she realized she had a tendency to simply bypass the formalities and focus the time and effort on what was of most importance to the mentee. She has had mentees assigned to her and has also had people reach out to her. Her experience in mentoring has led her to believe that mentoring happens everywhere, all the time and using all forms of communication.

Eve and Maria believe that people of all ages can provide guidance. Younger people can mentor older ones and vice versa.

Some of Eve's important mentors have been colleagues, who were about the same age as her.

The most important mentors for Maria Angela come from outside her corporate environment, and she finds comfort in addressing her issues with those external people rather than those embedded in her organization.

Some Names to Thank in Mentoring and Sponsoring

Maria Angela had mentors of pivotal importance, who later in her career were also sponsors for her. She would like to cite Dr. Michael L. Batzle, Dr. Thomas Davis, Anna Shaughnessy, Dr. Giuseppe Giannetto, and Susan Howes. Notably, the majority are not from the sector where she works, but from the academic sector. Maria's insertion in Middle Eastern culture and business style in oil and gas was greatly facilitated by Sid Williams and Vispi Dumasia, of Halliburton. Her colleagues in Kuwait Oil Company through the years have provided constant mentoring in short sessions, in which a two-way mentoring process occurs in a spontaneous way.

As an anecdote related to reverse or mutual mentoring, Maria Angela was preparing some Team Leaders of Kuwait Oil Company (KOC) for interviews at a selection for a Harvard Business program, and she highlighted the importance of looking people in the eye, to raise trust and express confidence. Then, one of the Team Leaders explained to her why this did not come easy to a Middle Eastern man from Kuwait: "*When I was a small kid, my parents would reprove me if I directly looked at them, warning me not to do that. And at school, kids would pick a fight with me, if I looked or stared at them, as it is considered a sign of open confrontation, inviting a response.*" This was a clear case of mutual mentoring, and it served Maria Angela to enhance her training, emphasizing the need of forgetting learned behaviors, to assume new ones. She was proud all KOC's Team Leaders applying to the Harvard program were accepted.

Another anecdote relates to overcoming the sense of modesty of Kuwaiti women leaders. For the mentoring of a KOC C-Suite woman, Maria Angela collected several photos of her in the KOC publications, to call her attention to the fact that she was always on the extreme lateral side of the photo, whereas she was the most important person in the portrayed groups. Maria Angela had to explain that it was necessary for her to literally "take the center spot" not for her, but for needed importance of her group of people, whom she represented, and also as a role model for all young, mid-career and even experienced women in KOC, as she was the only one in such high rank. Things changed, and after this mentoring session, Maria Angela was pleased to see this woman leader in the center of all photos.

Eve's best friend at Mobil, Nizar Djabbarah, was a fabulous sounding board for her. Contrary to some people's preconceived assumptions, some of Eve's best supporters, like Nizar and Aziz Odeh have been men of Middle Eastern origin.

For those who belong to a minority group in their organization, having a mentor from the same group can be important. At Chevron, Eve was lucky to have a good friend, June Gidman, who was about her age and level. They would walk together at lunchtime and talk about their issues, in what was a mutual mentoring process.

As an anecdote, one day, out of the blue, one of the highest-ranking women in the organization lashed out at Eve for no reason. When she spoke to her friend June about this leader, June revealed all the ways that woman had made her life miserable and the problems she had caused for other women at their level. That high-ranking woman was so insecure, she attacked every woman she thought might in any way be competition. She was a vicious queen bee.

Maria Angela experienced constantly this kind of female aggression against other talented women in Venezuela, USA and Kuwait, and at times she noticed their dedicated efforts to limiti her visibility, and prevent her from receiving promotions and relevant appointments.

Our Sponsoring

Maria Angela has lacked sponsorship in many phases of her career. A career in which she not only abruptly shifted settings, but also disciplines and cultures. Nevertheless, sponsors facilitated key steps to relevant roles, especially in professional societies. Susan Howes, Eve Sprunt, Kamel Ben Naceur, and Behrooz Fattahi have supported Maria Angela for leadership roles in the Society of Petroleum Engineers (SPE). Anna Shaughnessy and Nancy House sponsored Maria Angela at Society of Exploration Geophysicists (SEG) and positioned her to lead key programs and global committees like SEAM and others. They encouraged her to stretch her volunteerism to have greater impact.

Maria Angela acknowledges the significant support of several C-suite executives in PDVSA and in Kuwait Oil Company, who propelled her career advancement and leadership in her corporate tenures.

Maria Angela has had other key sponsors, she thinks are worth mentioning, including Madre Julia Pueyo, who endorsed her to become President of the TPA, Teachers and Parents Association of a major school in Venezuela, triggering a remarkable leadership experience, at the time of the very first protests from the education sector against the government in Venezuela. Also, Dr. Giuseppe Giannetto, rector of the Universidad Central de Venezuela, who endorsed her as a co-founder and the first Executive Director of that university alumni association. She also recognizes in her husband, Dr. Herminio Passalacqua, a constant mentor, able to separate an objective vision from their close relationship, and who has acted as a key mentor in her peak difficult work and life episodes.

Maria Angela thinks women seeking sponsors benefit from:

- Cultivating their network with information about their achievements
- Enhancing their professional branding
- Volunteering for leadership roles
- Asking potential sponsors for their support
- Understanding the formal but especially the informal hierarchies of power of their organizations
- Identifying potential sponsors both within and outside their organizations
- Building a strong network within your minority group
- Following and connecting with key women role models.

She emphazises that women need to recognize that hurdles created by jealousy, prima-donna complex, and egoism unfortunately may be more frequent among women, because they are self-conscious they are a minority, and sharing the throne is not easy. It would be healthy to ensure your sponsors include men and women at pair.

Key sponsors have also shaped Eve's career. At the beginning of her third year at MIT, Eve wanted a summer job and approched Prof. Gene Simmons, who contacted a friend of his, who worked at Shell Research and Development in Houston. That summer job altered the course of her life. Instead of targeting a career as a college professor, she changed her goal to wanting to be a research scientist at an oil company research lab, because she thought that being a working mother would be less of a problem as a research scientist than as a tenure-track professor.

In those days with very limited fertility treatments, a woman who wanted children had a narrower time window to give birth, which coincided with the time during which it was necessary to qualify for tenure. Eve didn't realize that in business, there is a less explicit timetable where early career achievements and advancement are as important as building up publications for an academic career. In both industry and academia women are challenged by having to build their reputations at the same time they are raising young children.

Eve's main way of sponsoring others has been by writing letters of recommendation and nominating people for various roles. She likes people to explain to her what they want to accomplish. Then, she asks questions to help them clarify their goals and find ways around the barriers they face. As a former SPE President, she is particularly surprised by how many people mention how a brief encounter with her influenced their life.

Eve's first important mentor was Prof. William F. Brace at MIT. When she started at MIT, she wanted to do research in seismology. Eve spent her sophomore year of college and the following summer gathering and analyzing data for a famous seismology professor, but when the work was published, she wasn't even mentioned in the acknowledgments. In comparison, a male student, the same age as Eve, co-authored a paper with that professor. Then Prof. Brace approached Eve, suggesting that she work with him instead of the seismology professor, and she seized the opportunity and co-authored three papers with him based on her undergraduate and master's thesis research.

The next pivotal sponsor for Eve was Dr. Aziz Odeh, the senior scientist at Mobil's Field Research Laboratory in Dallas, where she worked. Without asking or informing her, he nominated Eve for a SPE committee. When oil prices crashed, the chair of the committee lost his job. Eve had never attended a committee meeting, when she ended up single-handedly putting together that committee's session for the Annual Meeting. After she thanked Aziz for nominating her, Nizar Djabbarah told Eve that she was the first person who had ever thanked Aziz for being volunteered for a SPE committee. Subsequently, Aziz continued to nominate her for roles of increasing responsibility within SPE, which enabled Eve to rapidly rise through the SPE volunteer ranks and become the first woman to serve on the SPE Board of Directors.

When Eve asked Aziz how she could get company permission for the SPE roles, he told her, "*Tell them you will do most of the work on your own time and do so. It will get easier with time.*" That was very tough advice for a woman with two young children. Furthermore, in the 1980s the ability to work remotely was extremely limited. Eve learned Aziz was right. It did get easier with time. Involvement with SPE and the leaders Eve met through SPE shaped her life.

Eve has had a variety of other sponsors. One critical one was the Chief Technology Officer of Chevron, Dr. Don Paul, who hired her and guided her career at Chevron until his retirement.

Another major sponsor for Eve was Andrew Young, who was the 2003 President of SPE. She served with him on the SPE Board in the early 1990s, but was very surprised when Andrew called her to ask if she had thought she would be able serve as SPE President. Following Andrew's advice, she didn't seek Chevron's permission to serve until she was nominated to be the SPE president for 2006. Andrew was correct that getting that essential approval was easier if you were already the chosen candidate. Serving as SPE President was a wonderful, life-altering experience for Eve.

One potential concern for women is a mentor or sponsor taking advantage of the situation to shift the relationship from purely professional to one involving physical contact. Eve never experienced this problem, but knows women who have.

Throughout her career, Eve made a practice of minimizing and often completely avoiding alcohol consumption at professional events. This put her in a better position to maintain control of situations. Her view was that it was easier to maintain the line at drinking nothing alcoholic than to refuse another drink. If you are sober, you are better able to use your wits to protect yourself.

While many women want to be free to wear whatever they wish, Eve also chose to dress conservatively, so as to project an image that she was not an easy target. She has had some interesting conversations over the years with male faculty discussing how difficult it was to maintain their professional demeanor when a young student was from their perspective provocatively dressed. Her husband would also talk about women in his office whose clothing gave him glimpses of body parts of interest.

Eve's advice is to draw your lines between work and play clearly and to maintain them so that relationships are purely professional. When you blur those lines, you may have a wonderful romance or you may have a career damaging disaster. The risk averse approach is to keep work strictly professional and to set the tone from the beginning.

Analyzing Mentoring and Sponsoring

M. A. Capello and E. Sprunt, *Mentoring and Sponsoring*,
https://doi.org/10.1007/978-3-030-59433-6_29

Compiled Learnings About Mentoring

Mentoring was not a term in vogue until the mid-1990s. Then, leading companies in every sector began to recognize that it was difficult to assess the potential of every employee in their large organization. As a result, high-potential individuals were not getting the experiences necessary to enhance their productivity. Frustrated ambitious employees were leaving because of an unwritten rule that to advance to the C-suite, you had to secure a managerial post before 40 years of age.

Mentoring can enhance employees' skills, vision, and career planning, and enable the company to identify talent within their own organization. Over time, mentoring programs proved to be valuable, and in many cases essential, when replacements were needed for experienced employees who had retired, quit, or not returned after being furloughed during a recession.

By the early 2010s, structured mentoring programs were the norm in many top companies. Mentoring quickly expanded. In addition, with the push for greater diversity in leadership, employers recognized that mentors were needed to accelerate training and to identify the exceptional talents of a wide range of people who did not necessarily have the same ethnicity, culture and gender as the current leadership.

Companies and organizations dedicated to facilitating the advancement of under-represented groups emerged to address the gaps in leadership. These groups aimed to empower, mentor and strategically position women in energy in the USA and in the Middle East, black managers in the mining businesses of African countries, people of Latino ethnicity in USA academia, and Arabic speaking leaders for refugee camps and aid program management in North Africa and the Middle East.

Building Blocks of Mentoring

We interviewed and captured the insights and perspectives of more than 24 leaders in a variety of industries from different countries. We combined the wisdom we learned from those leaders with our own personal experience to identify key success factors for mentoring.

- Goal Setting for mentoring sessions
- Advice on short-term alternatives
- Long-term career advice and planning
- Sharing of the mentor's experience

- Probing questioning by the mentor
- Mentee listening and willing to explore alternatives.

Compiled Learnings About Sponsoring

When we started writing this book, we wanted to compile and showcase the importance of mentoring, but very quickly realized that a critical success factor for helping employees advance was sponsorship. From our perspective, sponsoring means that an individual with the right connections or in a role of power provides opportunities to another person in either his/her own organization or through his/her connections in another organization.

Sponsoring is a complex process. It entails risk for the endorser, whose reputation will be tarnished if the sponsored person does not perform well. Not all potential sponsors are willing to take the risk of openly endorsing someone, if they are unsure of a good outcome.

The sponsor provides opportunities, literally unlocking doors (and surmounting barriers) for the sponsored person. The opportunity may be a higher role, a project, a membership in an important committee, or a new job. The recommendation from a sponsor can tilt the balance in favor of a person. The more important the sponsor, the more weight his or her endorsement can have on the ultimate decision. Each door that is opened leads to a path to the door of another higher opportunity.

High performers, besides having to amass important and visible achievements, should cultivate multiple sponsors. Life has many twists and turns and even the best and most reliable sponsor may not be able to provide the all-important support in the future. Your sponsor may retire, become incapacitated or in some other way become incapable of providing future support. We have witnessed people whose careers have stalled or crumbled when their all-important sponsor ceases to be able to provide further support for them.

Almost all organizations have a pyramidal shape, in which it becomes increasingly more difficult to move to the next level. Only one person can rise to the top. At the lowest career levels, sponsorship may not be essential for a promotion. As someone progresses to higher levels, sponsorship becomes more and more essential.

Elizabeth Coffey, one of our interviewed leaders, informed us that optimally seven sponsors are necessary to guarantee smooth upward progression throughout one's career. People are constantly changing roles and employers in the corporate ecosystem, so there is a large risk of losing a critical sponsor.

Ambitious people must cultivate and constantly renew their pool of sponsors, to facilitate endorsements for smooth advancement.

Insights we gained from leaders in Africa and Latin America helped us understand why strict processes had to be put into action to ensure fairness in career succession decisions and to minimize opportunities for abuse and corrupt favoritism. The difference between sponsorship of deserving talent and corrupt cronyism and favoritism can be in the eyes of the beholders.

The practice of recommendations and sponsorship can be highly problematical when applied across national and/or cultural borders. For example, men in Asia or the Middle East may think it inappropriate to be a sponsor of women. In some areas, family or tribal names may be more important than talent or achievements. Modern companies in the energy sector in the Middle East have implemented several filter systems to ensure the application of fair processes to minimize the lack of objectivity in the ranking of personnel for promotion decisions.

A person may gain a sponsor through modeling exemplary "esprit du corps." In the oil and gas industry and in the restaurant sectors, we found examples where career progression favored individuals with a "corporate-driven" attitude and behavior. Those who act as if the company was his or her own enterprise and who visibly showcase their allegiance will be favored over those who modestly fulfill their goals.

Even though it is well-known that diversity spurs higher productivity and better decisions, only strong leaders are confident enough to surround themselves with a diverse team of people who will challenge their decisions and enrich their perspectives. Ideas that are challenged and tested are more likely to produce excellent financial results. True leaders recognize the value of a broad range of advisors as opposed to a team of agreeable clones.

Building Blocks of Sponsoring

Sponsoring requires navigation of a very fine line. Empathy between sponsor and sponsored is vital. Other critical building blocks include:

- Ambition
- Talent
- Strategic Thinking
- Motivation

Mentoring and Sponsoring Change with Experience

After analyzing the building blocks of mentoring and sponsoring, we realized it would be useful to provide guidance on where to focus the attention during your career evolution. Our goal is to support your search for the best mentors and sponsors for you, by enhancing your understanding of the key factors required to optimize the value of the precious time that mentors and sponsors are willing to spend to help you shape your career.

We focus on three stages of career development:

(1) ***Young Professionals (YP)***: Fresh graduates and new hires with up to ten years of experience. Typical roles in the energy sector would be Production Engineer, Accountant, and Researcher.
(2) ***Mid-Career (MC)***: Professionals with ten or more years of experience but usually less than 25 years of work experience and under 50 years of age, who have not yet reached a managerial (administrative career ladder) or regional-expert level (technical career ladder). In the energy sector typical roles would be Senior Engineer, Project Manager, Sr. Researcher, Section Head of Gathering Center; in the education sector, Associate Professor; in other sectors, examples include Sous Chef or Choir Director Assistant.
(3) ***Leaders***: Professionals with managerial or senior management roles, in administrative or technical positions who steer teams or programs with ample scope, generally more than 25 years of experience at work, and over 50 years of age. Typical roles in the energy sector are Field Development Manager, Operations Manager, Deputy CEO, Principal Researcher, Fellow, Asset Manager, CEO. In academia, Rector, Provost, Dean. In other sectors, Executive Chef, CEO, Orchestra Director, Artistic Director.

We focused our attention on key elements that would optimize value received from the mentoring and sponsoring to create our "CRITICAL LISTENING" and "BET" (Building priorities, Equal priorities and Tactics) models.

The Critical Listening Model

For mentoring, we focused on how the mentees should use the precious time in the mentoring sessions. Critical listening is of paramount importance at all stages of career evolution. Hence, this is called "THE CRITICAL LISTENING MODEL".

The elements that compose this model are

1. Smart Questioning
2. Prioritizing Preferences and Personal Issues
3. Listening

Each of these three elements are important all along the career path. However, the focus of mentoring sessions and division of "airtime" should vary with career maturity.

1. Smart Questioning

Prior to a mentoring session, you should do your homework and be ready to maximize the precious time with your mentor. Ask about long-term career strategies including the importance of additional training, certification and different career paths. Use your questions to focus the discussion on the issues of greatest importance to you.

2. Prioritize Issues and Preferences

Before the mentoring session think about your priorities and any concerns or limitations you have. Develop a concise "elevator speech" that explains to the mentor your goals and accomplishments. This "elevator speech" should be five minutes or less; it's a pitch that lasts the 3–5 min it takes to go from ground floor to the executive suite in an elevator if you encounter a senior leader in your organization. You want to focus the discussion, but you are there to get the mentor's advice on the issues of concern to you, not to hear yourself speak. For example, you may want to get your mentor's advice on a technical versus managerial career path and/or on various work-life issues.

3. Listening

To benefit from a mentoring session, you must be a good listener. Treat each mentoring session as if there you will not have the opportunity to talk to this mentor again. The time of the mentor is indeed very precious, so every moment of active listening counts.

Listen carefully to grasp the mentor's essential messages. The guidance may be conveyed by a combination of body language, stories and recommendations. Even if you do not agree with what you are hearing, listen closely, because you may find some of the guidance to be extremely useful.

Be prepared to take notes about references to papers, people to contact, emails and telephone numbers, as these may pop up in the conversation. Don't try to write everything down, just the highlights. Listening carefully to understand and capture the mentor's main advice.

A good mentor will also ask questions to trigger your thoughts about the way forward and to understand your preferences. You need to be prepared to listen carefully to understand what the mentor wants to know to steer the conversation to address the most important issues for you.

Mentee's Use of Time

We estimated the importance of each of the three factors of mentee's use of time along career paths, based on the feedback we received from the twenty leaders we interviewed, and our own experience combined with the insights we gained from the young professionals we interviewed (see Table 1). This information on Percentage of Mentees Use of Time for Young Professionals (YP), Mid-Career (MC) professionals and Leaders is displayed in table and pie chart format below.

In pie chart graphical format (Fig. 2), this information looks more meaningful, showcasing that at every stage, critical listening is essential for successful mentoring.

Table 1 Mentees use of time in percentages

Mentees use of time			
Activity during mentoring session	YP (% of time)	MC (% of time)	Leader (% of time)
Smart questioning	5	15	40
Prioritize issues and preferences	15	25	5
Listening	**80**	**60**	**55**
Total	100	100	100

The "Critical Listening" Model – The Use of Mentoring Time by Mentees

A mentee uses a mentoring session with different focus depending on how experienced they are in their career paths. The three pie charts below show this evolution.

1. **Young Professionals**: Most of the mentoring sessions time is dedicated to listening to what the mentor explains about future outlooks or options.
2. **Mid-Career**: Prioritizing. Conversations about how to prioritize their activities with regard to future career options and outlooks
3. **Leaders or Experienced Professionals** – Asking their mentors about key areas.

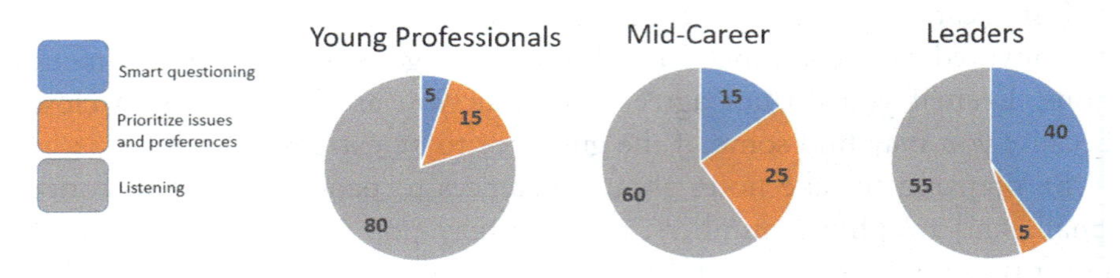

Fig. 2 The Critical listening model, establishing the use of Mentoring time by professionals during their careers: YP (Young Professionals; MC (Mid-Career); and Leaders show a shift in the distribution of their time in the activities of Smart Questioning, Prioritization/Preferences, and Listening, as shown in the three pie charts

The BET Model

We also identified three factors that are essential for identifying and securing sponsors, who can provide critical career acceleration. Our "BET" (Building priorities, Equal priorities and Tactics) model highlights the three aspects necessary to secure important sponsors. As we have discussed, a single sponsor can make a huge difference, but if you want to maximize your chance of success, you should have multiple sponsors.

The elements that compose our model are:

1. Goal
2. Building a Network
3. Tactics to find a Sponsor

However, as difficult as it is to find good mentors, it takes even more effort to secure meaningful sponsors. Following our model will enable you to take a more strategic approach to securing multiple sponsors. As with mentoring, the time we should spend on the different key components depends on where we are in our career.

1. Goals

Many of the leaders we interviewed ask their mentees to come with at least some thought applied to where they want to be at various points in their career. Life happens and many things may change our priorities, but when we are setting goals, we need to think five, ten and twenty years into the future.

One way to do this is to look at the inverse problem. Ask yourself, "Where do I want to be at the end my career?" Then figure out where you should be at various milestones to have any chance of achieving your goal. You must be honest with yourself on what trade-offs you are willing to make to reach your goal.

In our model, we show that from early career through the mid-point of your career you will need to spend more time thinking about your long-term goals and who might sponsor you. We have set the amount of time for thinking about your career goals to be almost equal for early and mid-career, because many people do a mid-career examination and decide to do something different. There can still be time to pivot and sponsors are very critical in making a major shift.

Once you have risen to a leadership role, the set of people who can sponsor you to rise further is much smaller. Top leaders often know what they want and have their eyes focused on specific opportunities.

2. Building a Network

Networking is one of the most important things that you can do to advance your career. Life is full of surprises, and you never know when you may need to reinvent yourself and hunt for new employment. Your network is your safety net. No matter how secure you think your job is, you should build a strong network not only within your organization, but within the greater community associated with your profession.

One of the best investments you can make in your future is getting involved with the important professional societies in your profession. We don't just mean paying your dues and attending meetings. You should be an active volunteer. When you serve as a volunteer, many people outside your current organization get to know your strengths and become familiar with your work ethic. This is a strategic way to build a robust personal network that can help you survive major downturns that impact not just your employer, but your entire industry. Think of the time you spend on professional society work as a major investment in your future and your financial security.

3. Tactics to find a Sponsor

As we discussed, it is much more difficult to find a sponsor than a mentor. Everyone can offer advice, but a much smaller number of people are able and willing to help us get a job. Some of those who could be valuable sponsors may not even realize it.

Instead of asking people to sponsor you for something, start the conversation by asking them for advice. Everyone can provide advice, but many people may not even realize that they can in some way help you get the job you want. By asking for advice (which is, in effect, mentoring), you are engaging them in a discussion and getting them interested in the challenges you are facing. As the discussion evolves, they may realize ways in which they can help (essentially sponsor) you. The sponsoring may come in the form of them calling someone to ask about opportunities. It may then be followed by them recommending you in one way or another for the position.

Make sure you pay attention to your professional branding, and that you communicate your achievements to key people in your organization. If you are not noticeable, your chances of gaining the attention of potential sponsors are scarce.

When you ask for advice you may find out about opportunities you hadn't considered that are even more attractive than what you originally were targeting.

If you are a leader, the top positions of interest to you are very difficult to secure. You need to spend a lot of time thinking about who can help you get the position and who might be able to introduce you to a key decision maker. These may be people connected to people who can help you. The higher the position you seek, the more time you need to invest in thinking how to leverage existing connections to meet people who can help you meet others. It turns into a complex chain of connections.

Time usage in Finding a Sponsor

We estimated the importance of each of the three factors individuals need to care about along their career paths (Table 2). We also estimated these based on the insights compiled in our interviews with leaders in different areas and our own experience to which we added the views we gained from the young professionals we interviewed.

Table 2 Use of Time to find Sponsors in YP (Young Professionals); MC (Mid-Career); and Leaders, for Goal setting, Building Network and Tactics at work

Finding sponsors use of time

Focus	YP (% of time)	MC (% of time)	Leader (% of time)
Goal	35	30	5
Building network	**55**	**45**	10
Tactics	10	25	**85**
Total	100	100	100

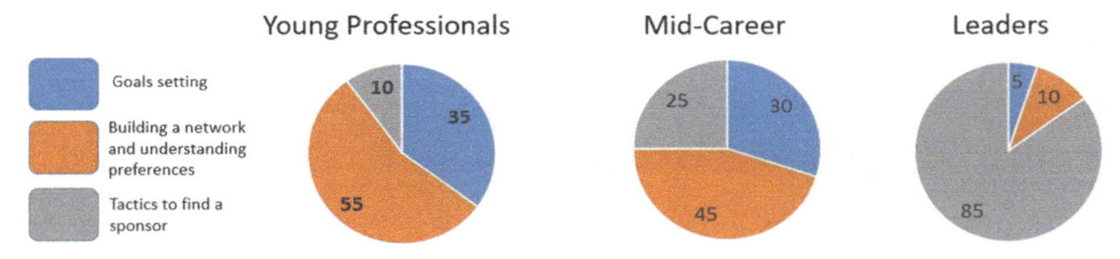

Fig. 3 The BET model implies an evolution of the distribution of time in finding sponsors for Goal setting, Building network/Preferences and Tactics, as done by YP (Young Professionals); MC (mid-Career); and Leaders

In graphical format (Fig. 3), the analysis shows the evolution of each of the factors. At the early and mid-career stages, building your network is of paramount importance. In contrast, those already in leadership positions have strong networks and need to focus on tactics. As a leader, they should already have an extensive network.

The importance of setting the goals is critical at the Young Professionals and mid-career stage, but the targets of leaders are narrower, so more of their time should be engaged in tactics rather than in setting their personal career goals and network building. A leader must spend a lot of time on his organization's goals as opposed to their personal career goals.

Why the Keys?

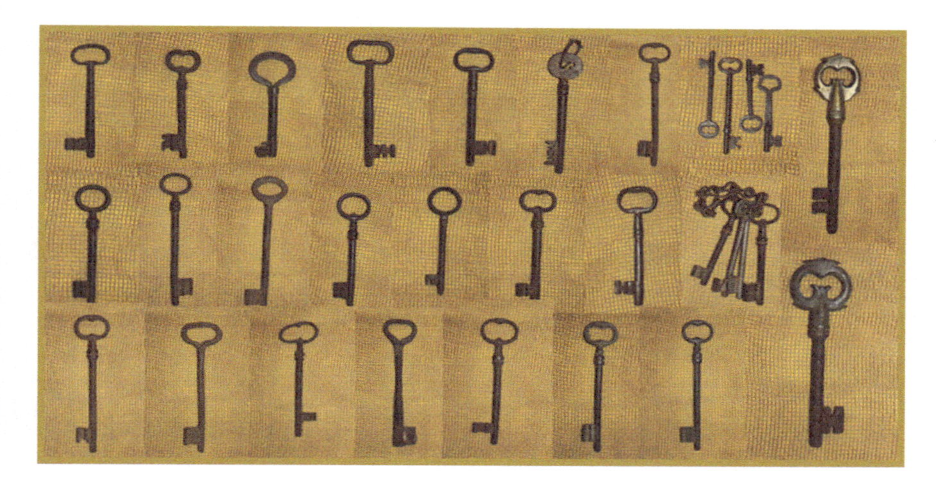

When we interviewed Her Highness Sheikha Intisar AlSabah, a Princess of the Kuwait Royal Family, she explained that sponsoring is like providing the keys to open doors. "I have the keys to Kuwait, but if I want to do things in other countries, I will have to look for someone who has the right keys."

This analogy inspired us to illustrate our book with photos of keys. Vintage and antique keys were selected because each one was forged and crafted individually.

Each key would then be unique, like our interviewees.

The photos were taken by Dr. Herminio Passalacqua, who besides being a pioneering geophysicist in electromagnetic methods and a dedicated mentor

M. A. Capello and E. Sprunt, *Mentoring and Sponsoring*,
https://doi.org/10.1007/978-3-030-59433-6

throughout his career, is a gifted photographer. Herminio used a portion of his quarantine time to photograph his collection of antique and vintage keys, selecting the most interesting ones for this project. An artist of photography, he is also Maria Angela's husband.

Printed in the United States
By Bookmasters